Springer Tracts in Mechanical Engineering

Springer Tracts in Mechanical Engineering (STME) publishes the latest developments in Mechanical Engineering - quickly, informally and with high quality. The intent is to cover all the main branches of mechanical engineering, both theoretical and applied, including:

- Engineering Design
- Machinery and Machine Elements
- Mechanical Structures and Stress Analysis
- Automotive Engineering
- Engine Technology
- Aerospace Technology and Astronautics
- Nanotechnology and Microengineering
- Control, Robotics, Mechatronics
- MEMS
- Theoretical and Applied Mechanics
- Dynamical Systems, Control
- Fluids Mechanics
- Engineering Thermodynamics, Heat and Mass Transfer
- Manufacturing
- Precision Engineering, Instrumentation, Measurement
- Materials Engineering
- Tribology and Surface Technology

Within the scope of the series are monographs, professional books or graduate textbooks, edited volumes as well as outstanding PhD theses and books purposely devoted to support education in mechanical engineering at graduate and post-graduate levels.

Indexed by SCOPUS, zbMATH, SCImago.

Please check our Lecture Notes in Mechanical Engineering at http://www.springer.com/series/11236 if you are interested in conference proceedings.

To submit a proposal or for further inquiries, please contact the Springer Editor **in your region**:

Ms. Ella Zhang (China)
Email: ella.zhang@springernature.com
Priya Vyas (India)
Email: priya.vyas@springer.com
Dr. Leontina Di Cecco (All other countries)
Email: leontina.dicecco@springer.com

All books published in the series are submitted for consideration in Web of Science.

More information about this series at https://link.springer.com/bookseries/11693

Dan B. Marghitu · Hamid Ghaednia · Jing Zhao

Mechanical Simulation with MATLAB®

 Springer

Dan B. Marghitu
Mechanical Engineering Department
Auburn University
Auburn, AL, USA

Hamid Ghaednia
Department of Orthopedic Surgery
Massachusetts General Hospital
Boston, MA, USA

Jing Zhao
Mechanical Engineering Department
Auburn University
Auburn, AL, USA

ISSN 2195-9862   ISSN 2195-9870 (electronic)
Springer Tracts in Mechanical Engineering
ISBN 978-3-030-88101-6   ISBN 978-3-030-88102-3 (eBook)
https://doi.org/10.1007/978-3-030-88102-3

This Springer imprint is published by the registered company Springer Nature Switzerland AG
The registered company address is: Gewerbestrasse 11, 6330 Cham, Switzerland

Preface

This book deals with the simulation of the mechanical behavior of engineering structures, mechanisms, and components. It presents a set of strategies and tools for formulating the mathematical equations and the methods of solving them using MATLAB. For the same mechanical systems, it also shows how to obtain solutions using different approaches. It then compares the results obtained with the two methods. By combining fundamentals of kinematics and dynamics of mechanisms with applications and different solutions in MATLAB of problems related to gears, cams, and multilink mechanisms, and by presenting the concepts in an accessible manner, this book is intended to assist advanced undergraduate and mechanical engineering graduate students in solving various kind of dynamical problems by using methods in MATLAB. It also offers a comprehensive, practice-oriented guide to mechanical engineers dealing with kinematics and dynamics of several mechanical systems.

Auburn, USA   Dan B. Marghitu
Boston, USA   Hamid Ghaednia
Auburn, USA   Jing Zhao

Contents

Chapter 1
Introduction

Abstract The structure of the mechanical systems is analyzed: links, joints, degrees of freedom of the joint, degrees of freedom, independent contours, dyads. For the planar mechanisms the contour diagram and the dyads are introduced. Formulas for kinematics and dynamics of the rigid body are presented.

1.1 Kinematic Pairs

Linkages are made up of links and joints and are basic elements of mechanisms and robots. A link (element or member) is a rigid body with nodes. The nodes are points at which links can be connected. Figure 1.1 shows a link with two nodes, a binary link. The links with three nodes are ternary links. A kinematic pair or a joint is the connection between two or more links. The kinematic pairs give relative motion between the joined elements. The degree of freedom of the kinematic pair is the number of independent coordinates that establishes the relative position of the joined links.

A joint has $(6 - i)$ degrees of freedom where i is the number of restricted relative movements. A planar one degree of freedom kinematic pair, c_5, removes 5 degrees of freedom and allows one degree of freedom. The planar two degrees of freedom kinematic pair, c_4, has two degrees of freedom and removes 4 degrees of freedom. To find the degrees of freedom of a kinematic pair one element is hold to be a reference link and the position of the other element is found with respect to the reference link. Figure 1.2a shows a slider (translational or prismatic) joint that allows one translation (T) degree of freedom between the elements 1 and 2. Figure 1.2b represents a rotating pin (rotational or revolute) joint that allows one rotational (R) degree of freedom between links 1 and 2. The slider and the pin joints are c_5 joints. The c_5 joints allow one degrees of freedom and is called full-joint. For the two degrees of freedom joints, c_4, there are two independent, relative motions, translation (T) and rotation (R), between the joined links. Two degrees of freedom joints are shown in Fig. 1.3. The two degrees of freedom joint is called half-joint and has 4 degrees of constraint. For a planar system there are two kinds of joints c_5 and c_4. A joystick (ball-and-socket joint, or a sphere joint) is a three degrees of freedom joint (3 degrees of constraint,

© The Author(s), under exclusive license to Springer Nature Switzerland AG 2022
D. B. Marghitu et al., *Mechanical Simulation with MATLAB®*,
Springer Tracts in Mechanical Engineering,
https://doi.org/10.1007/978-3-030-88102-3_1

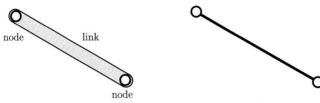

Fig. 1.1 Link with two nodes

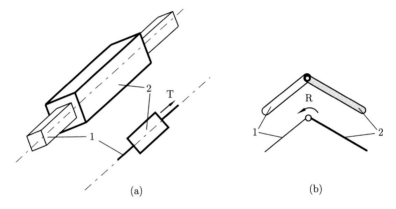

Fig. 1.2 One degree of freedom joints, c_5: **a** slider joint and **b** pin joint

Fig. 1.3 Two degrees of freedom joints, c_4: **a** two curves in contact and **b** cam-follower joint

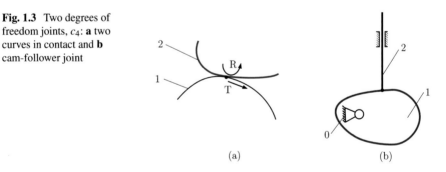

c_3) and allows three independent motions. An example of a four degrees of freedom joint, c_2, is a cylinder on a plane. A five degrees of freedom joint, c_1, is represented by a sphere on a plane. The contact between the links can be a point, a curve, or a surface. A point or curve contact defines a higher joint and a surface contact defines a lower joint. The order of a joint is defined as the number of links joined minus one. Two connected links have the order one (one joint) and three connected links have the order two (two joints).

1.2 Degrees of Freedom

The number of independent variables that uniquely defines the position of a mechanical system in space at any time is defined as the number of degrees of freedom (DOF). The number of DOF is stated about a reference frame.

Figure 1.4 represents a rigid body (RB) moving on xy-plane. The distance between any two particles on a rigid body is constant at any time. Three DOF are needed to define the position of a free rigid body in planar motion: two linear coordinates (x, y) to define the position of a point on the rigid body, and one angular coordinate (θ) to define the angle of the body with respect to the reference axes. The particular selection of the independent measurements to define its position is not unique. A free rigid body moving in a three-dimensional (3-D) space has six DOF: three lengths (x, y, z), and three angles $(\theta_x, \theta_y, \theta_z)$. Next only the two-dimensional motion will be presented. A rigid body in planar motion has pure rotation, if the body possesses one point (center of rotation) that has no motion with respect to a fixed reference frame. The points on the body describe arcs with respect to its center. A rigid body in planar motion has pure translation if all points on the body describe parallel paths. A rigid body in planar motion has complex or general plane motion if it has a simultaneous combination of rotation and translation. The points on the body in general plane motion describe non-parallel paths at an instantaneous center of rotation will change its position.

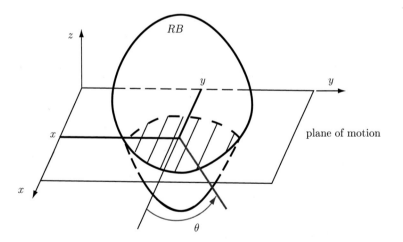

Fig. 1.4 Free rigid body in planar motion with three DOF

1.3   Kinematic Chains

Bodies linked by joints form a kinematic chain as shown in Fig. 1.5. A contour or loop is a configuration described by a closed polygonal chain consisting of links connected by joints, Fig. 1.5a. The closed kinematic chains have each link and each joint incorporated in at least one loop. The closed loop kinematic chain in Fig. 1.5a is defined by the links 0, 1, 2, 3, and 0. The open kinematic chain in Fig. 1.5b is defined by the links 0, 1, 2, and 3. A mechanism is a closed kinematic chain. A robot is an open kinematic chain. The mixed kinematic chains are a combination of closed and open kinematic chains. The crank is a link that has a complete revolution about a fixed pivot. Link 1 in Fig. 1.5a is a crank. The rocker is a link with oscillatory rotation and is fixed to the ground. The coupler or connecting rod is a link that has complex motion and is not fixed to the ground. Link 2 in Fig. 1.5a is a coupler. The ground or the fixed frame is a link that is fixed (non-moving) with respect to the reference frame. The ground is denoted with 0.

A planar mechanism is shown in Fig. 1.6a. The mechanism has five moving links 1, 2, 3, 4, 5, and a fixed link, the ground 0. The translation along the i axis is denoted by T_i, and the rotation about the i axis is denoted by R_i, where $i = x, y, z$. The motion of each link in the mechanism is analyzed in terms of its translation and rotation about the fixed reference frame xyz. The link 0 (ground) has no translations and no rotations. The link 1 has a rotation motion about the z axis, R_z. The link 2 has a planar motion (xy is the plane of motion) with a translation along the x axis, T_x, a translation along the y axis, T_y, and a rotation about the z axis, R_z. The link 3 (slider) has a rotation motion about the z axis, R_z. The link 4 has a planar motion (xy the plane of motion) with a translation along x, T_x, a translation along y, T_y, and a rotation about z, R_z. The link 5 has a rotation about the z axis, R_z.

A graphical construction for the mechanism connectivity is the contour diagram [3, 18]. The numbered links are the nodes of the diagram and are represented by

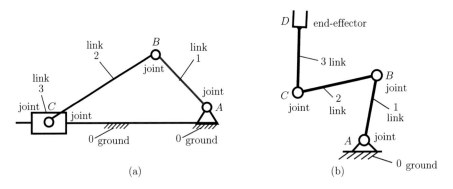

(a)   (b)

Fig. 1.5 Kinematic chains: **a** closed kinematic chain, mechanism and **b** open kinematic chain, robot

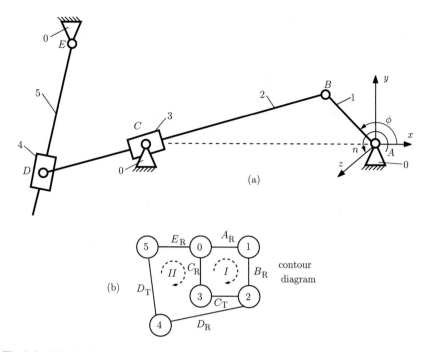

Fig. 1.6 **a** Mechanism with five moving links and **b** contour diagram

circles, and the joints are represented by lines that connect the nodes. Figure 1.6b is the contour diagram of the planar mechanism. The link 1 is connected to ground 0 at A and to link 2 at B with revolute joints. The link 2 is connected to link 3 at C with a slider joint. The link 3 is connected to ground 0 at C with a revolute joint. Next, the link 2 is connected to link 4 at D with a revolute joint. Link 2 is a ternary link because it is connected to three links. Link 4 is connected to link 5 at D with a slider joint. Link 5 is connected to ground 0 at E with a revolute joint. The independent contour is the contour with at least one link that is not included in any other contours of the chain. The number of independent contours, N, of a kinematic chain is computed as

$$N = c - n, \tag{1.1}$$

where c is the number of joints, and n is the number of moving links.

For the mechanism shown in Fig. 1.6a the independent contours are $N = c - n = 7 - 5 = 2$, where $c = 7$ is the number of joints and $n = 5$ is the number of moving links. Some contours of the mechanisms can be selected as: 0-1-2-3-0, 0-1-2-4-5-0, and 0-3-2-4-5-0. Only two contours are independent contours.

The number of degrees of freedom or the mobility of a mechanical system is the number of independent coordinates required to define the position of the system in space and time. For mechanisms with planar motion the number of degrees of freedom is

$$M = 3\,n - 2\,c_5 - c_4, \tag{1.2}$$

where n is the number of moving links, c_5 is the number of one degree of freedom joints, and c_4 is the number of two degrees of freedom joints. For the mechanism shown in Fig. 1.6a there are 5 moving links and 7 one degree of freedom joints. The number of degrees of freedom is

$$M = 3\,n - 2\,c_5 - c_4 = 3\,(5) - 2\,(7) - 0 = 1.$$

The system group or the fundamental kinematic chain is a kinematic chain that can be connected or disconnected from a mechanism at the same time preserving the original degrees of freedom of the system. The mechanism is structurally modified and the developed linkage has the same degrees of freedom. Equation (1.2) is used to obtain a kinematic chain with zero DOF

$$3\,n - 2\,c_5 = 0. \tag{1.3}$$

For planar mechanisms, a two degrees of freedom joints, c_4, can be substituted with two one degree of freedom joints, c_5, and an extra link [47].

1.4 Type of Dyads

The dyad (binary group) is a fundamental kinematic chain with two links ($n = 2$) and three one degree of freedom joints ($c_5 = 3$). Figure 1.7 depicts different types of dyads: rotation rotation rotation (dyad RRR) or dyad of type one, Fig. 1.7a; rotation rotation translation (dyad RRT) or dyad of type two, Fig. 1.7b; rotation translation rotation (dyad RTR) or dyad of type three, Fig. 1.7c; translation rotation translation (dyad TRT) or dyad of type four, Fig. 1.7d; rotation translation translation (dyad RTT) or dyad of type five, Fig. 1.7e. The advantage of the group classification of a mechanical system is in its simplicity. The solution of the whole mechanical system can be obtained by adding partial solutions of different fundamental kinematic chains [55–57].

The number of DOF for the mechanism in Fig. 1.8 is $M = 1$. If $M = 1$, there is one driver link (one actuator). The rotational link 1 can be selected as the driver link. If the driver link is separated from the mechanism the remaining moving kinematic chain (links 2, 3, 4, 5) has the number of DOF equal to zero. The dyad is the simplest system group with two links and three joints. On the contour diagram, the links 2 and 3 form a dyad and the links 4 and 5 represent another dyad. The mechanism has been decomposed into a driver link (link 1) and two dyads (links 2 and 3, and links 4 and

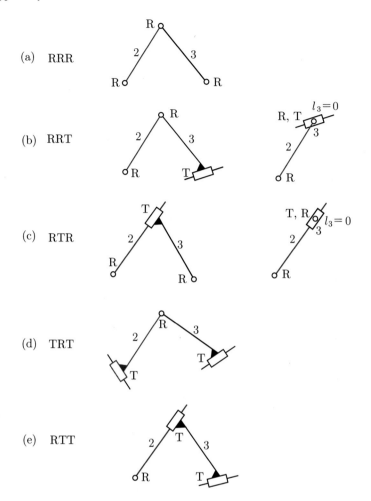

Fig. 1.7 Types of dyads: **a** RRR, **b** RRT, **c** RTR, **d** TRT, and **e** RTT

5). The dyad with the links 2 and 3 is a RTR dyad and the dyad with the links 4 and 5 is also a RTR dyad. The whole mechanism can be symbolized as a R-RTR-RTR mechanism.

For planar mechanisms the two degrees of freedom joints can be substituted and mechanisms with one degree of freedom joints are obtained. The transformed mechanism has to be equivalent with the initial mechanism from a kinematical point of view. The number of degrees of freedom of the transformed mechanism has to be equal to the number of degrees of freedom of the initial mechanism. The relative motion of the links of the transformed mechanism has to be the same as the relative motion of the links of the initial mechanism.

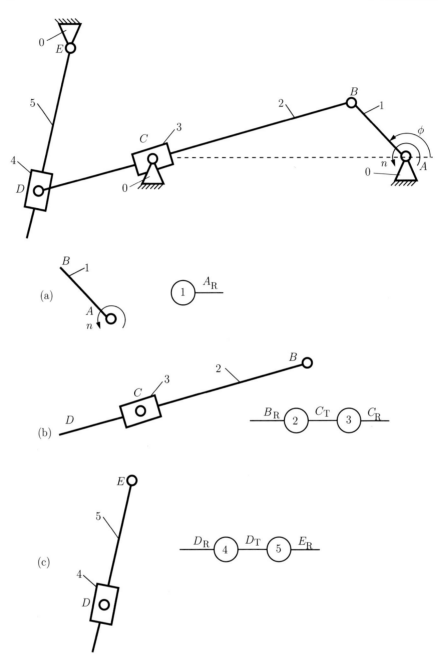

Fig. 1.8 R-RTR-RTR Mechanism: **a** driver link R, **b** dyad RTR, and **c** dyad RTR

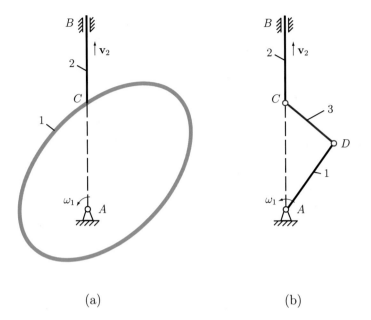

Fig. 1.9   a Cam and follower mechanism and **b** equivalent R-RRT mechanism

A two degrees of freedom joint can be substituted with one link ($n = 1$) and two one degree of freedom joints ($c_5 = 2$).

Figure 1.9 shows a cam and follower mechanism. There is a two degrees of freedom joint at the contact point C between the links 1 and 2. The two degrees of freedom joint at C, is replaced with one link, link 3, and two one degree of freedom joints at C and D as shown in Fig. 1.9b. To have the same relative motion, the length of link 3 has to be equal to the radius of curvature ρ of the cam at the contact point C.

In this way the two degrees of freedom joint at the contact point can be substituted with two one degree of freedom joints, C and D, and an extra link 3, between links 1 and 2. The new mechanism still has one degree of freedom, and the cam and follower system (0, 1, and 2) is a R-RRT mechanism (0, 1, 2, and 3) in a different aspect.

1.5   Position Analysis for Links

A planar straight link with the end nodes at A and B is considered. The coordinates of the joint A are $(x_A, \, y_A)$ and the coordinates of the joint B are $(x_B, \, y_B)$. The length $AB = l_{AB}$ is constant and the following relation can be written

$$(x_B - x_A)^2 + (y_B - y_A)^2 = AB^2, \tag{1.4}$$

The angle of the link AB with the horizontal axis Ox is ϕ and the slope m of AB is

$$m = \tan \phi = \frac{y_B - y_A}{x_B - x_A}. \tag{1.5}$$

The equation of the straight link is $y = m\,x + n$, where x and y are the coordinates of any point on this link and n is the intercept of AB with the vertical axis Oy.

The R-RRT (slider-crank) mechanism shown in Fig. 1.10a has the dimensions: $AB = 0.5$ m and $BC = 1$ m. The driver link 1 makes an angle $\phi = \phi_1 = 60°$ with the horizontal axis. The positions of the joints and the angles of the links with the horizontal axis will be calculated. A Cartesian reference frame xy is selected. The joint A is in the origin of the reference frame, $A \equiv O$,

$$x_A = 0, \quad y_A = 0.$$

The coordinates of the joint B are

$$x_B = AB \, \cos \phi = (0.5) \, \cos 60° = 0.250 \text{ m},$$
$$y_B = AB \, \sin \phi = (0.5) \, \sin 60° = 0.433 \text{ m}. \tag{1.6}$$

The coordinates of the joint C are x_C and y_C. The joint C is located on the horizontal axis $y_C = 0$. The length of the segment BC is constant

$$(x_B - x_C)^2 + (y_B - y_C)^2 = BC^2, \tag{1.7}$$

or

$$(0.250 - x_C)^2 + (0.433 - 0)^2 = 1^2.$$

The two solutions for x_C are:

$$x_{C_1} = 1.151 \text{ m} \quad \text{and} \quad x_{C_2} = -0.651 \text{ m}.$$

To determine the position of the joint C, for the given angle of the link 1, an additional condition is needed. For the first quadrant, $0 \le \phi \le 90°$, one condition can be $x_C > x_B$. The x-coordinate of the joint C is $x_C = x_{C_1} = 1.151$ m. The angle of the link 2 (link BC) with the horizontal is

$$\phi_2 = \arctan \frac{y_B - y_C}{x_B - x_C}.$$

The numerical solutions for ϕ_2 is 154.341 degrees. The MATLAB code is:

```
clear; clc; close all;
% Input data
AB = 0.5; % (m)
```

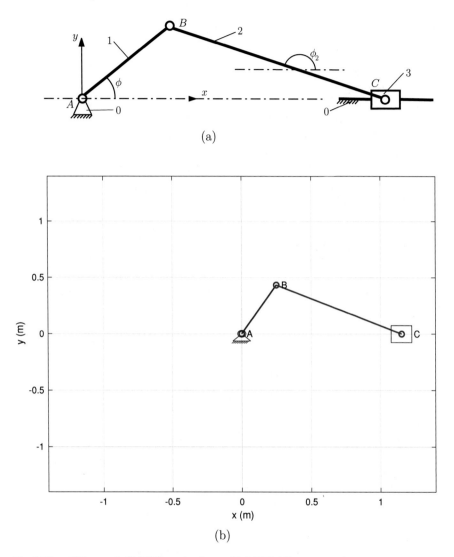

Fig. 1.10 **a** Slider-crank (R-RRT) mechanism and **b** MATLAB representation

```
BC = 1;    % (m)
phi = pi/3; % (rad)

% position of joint A (origin)
xA = 0; yA = 0;

% position of joint B
xB = AB*cos(phi);
```

```
yB = AB*sin(phi);

% position of joint C
yC = 0;
% xC unknown
syms xCsol
% distance formula: BC=constant
eqnC = ( xB - xCsol )^2 + ( yB - yC )^2 - BC^2;
% solve equation eqnC
solC_ = solve(eqnC, xCsol);
% two solutions for xC
% first  component of the vector solC
xC1 = eval(solC_(1))
%    -0.6514
% second component of the vector solC
xC2 = eval(solC_(2))
%     1.1514

% select the correct position for C
% or the given input angle
if xC1 > xB xC = xC1; else xC = xC2; end

% angle of the link 2 with the horizontal
phi2 = atan((yB-yC)/(xB-xC));

fprintf('Results \n\n')
% Print the coordinates of B
fprintf('xB =   %6.3f (m)\n', xB)
fprintf('yB =   %6.3f (m)\n', yB)
% Print the coordinates of C
fprintf('xC =   %6.3f (m)\n', xC)
fprintf('yC =   %6.3f (m)\n', yC)
% Print the angle phi2
fprintf('phi2 = %6.3f (degrees) \n',180+phi2*180/pi);
```

A four-bar (R-RRR) planar mechanism is shown in Fig. 1.11. The driver link is the rigid link 1 (the element AB) and the origin of the reference frame is at A. The following data are given: $AB = 0.15\,\text{m}$, link 2 $BC = 0.35\,\text{m}$, link 3 $CD = 0.25\,\text{m}$, $x_D = 0.30\,\text{m}$, and $y_D = 0\,\text{m}$. The angle of the driver link 1 with the horizontal axis is $\phi = \phi_1 = 45°$. Next the positions of the joints and the angles of the links with the horizontal axis are calculated.

A Cartesian reference frame xyz with the unit vectors $[\imath, \jmath, k]$ is selected. Since the joint A is the origin of the reference system $A \equiv O$ the coordinates of A are $x_A = 0$, $y_A = 0$.

The coordinates of the joint B, x_B and y_B are

Fig. 1.11 Four-bar (R-RRR) mechanism

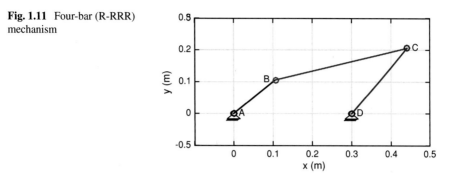

$$x_B = AB \cos \phi = 0.106 \text{ m}, \quad y_B = AB \sin \phi = 0.106 \text{ m}.$$

The position vector of B is $\mathbf{r}_B = x_B \mathbf{1} + y_B \mathbf{J}$. The unknowns are the coordinates of the joint C, x_C and y_C. Knowing the positions of the joints B and D, the position of the joint C is computed using the fact that the lengths of the links BC and CD are constants

$$(x_C - x_B)^2 + (y_C - y_B)^2 = BC^2,$$
$$(x_C - x_D)^2 + (y_C - y_D)^2 = CD^2,$$

or

$$(x_C - 0.106)^2 + (y_C - 0.106)^2 = 0.35^2,$$
$$(x_C - 0.300)^2 + (y_C - 0)^2 = 0.25^2. \tag{1.8}$$

Solving the system of equations, Eqs. 1.8, two sets of solutions are found for the position of the joint C. These solutions are

$$x_{C_1} = 0.2028 \text{ m}, \quad y_{C_1} = -0.2303 \text{ m} \quad \text{and} \quad x_{C_2} = 0.4415 \text{ m}, \quad y_{C_2} = 0.2061 \text{ m}.$$

To determine the correct position of the joint C for this mechanism, a constraint condition is needed: $y_C > 0$. The coordinates of joint C have the following numerical values

$$x_C = 0.441 \text{ m} \quad \text{and} \quad y_C = 0.206 \text{ m}.$$

The angles of the links 2 and 3 with the horizontal are

$$\phi_2 = \arctan \frac{y_B - y_C}{x_B - x_C}, \quad \phi_3 = \arctan \frac{y_D - y_C}{x_D - x_C}.$$

The results are $\phi_2 = 16.613°$ and $\phi_3 = 55.540°$. The MATLAB code for the position analysis is:

```
% Input data
AB=0.15; %(m)
BC=0.35; %(m)
CD=0.25; %(m)
xD=0.30; %(m)
yD=0; %(m)
phi = pi/4 ; %(rad)

xA = 0; yA = 0;
rA_ = [xA yA 0];
rD_ = [xD yD 0];

xB = AB*cos(phi);
yB = AB*sin(phi);
rB_ = [xB yB 0];

% Position of joint C
% Distance formula: BC=constant
syms xCsol yCsol
eqnC1 = (xCsol - xB )^2 + ( yCsol - yB )^2 - BC^2;
% Distance formula: CD=constant
eqnC2 = (xCsol - xD )^2 + ( yCsol - yD )^2 - CD^2;

% Simultaneously solve above equations
solC_ = solve(eqnC1, eqnC2, xCsol, yCsol);
% Two solutions for xC - vector form
xCpositions = eval(solC_.xCsol);
% Two solutions for yC - vector form
yCpositions = eval(solC_.yCsol);
% Separate the solutions in scalar form
% first  component of the vector xCpositions
xC1 = xCpositions(1);
% second component of the vector xCpositions
xC2 = xCpositions(2);
% first  component of the vector yCpositions
yC1 = yCpositions(1);
% second component of the vector yCpositions
yC2 = yCpositions(2);

% select the correct position for C
% for the given input angle
```

```
if xC1 > xD
    xC = xC1; yC=yC1;
else
    xC = xC2; yC=yC2;
end
rC_ = [xC yC 0]; % Position vector of C

% Angles of the links with the horizontal
phi2 = atan((yB-yC)/(xB-xC));
phi3 = atan((yD-yC)/(xD-xC));

fprintf('Results \n\n')
fprintf('rA_  =   [ %6.3f, %6.3f, %g ] (m)\n', rA_)
fprintf('rD_  =   [ %6.3f, %6.3f, %g ] (m)\n', rD_)
fprintf('rB_  =   [ %6.3f, %6.3f, %g ] (m)\n', rB_)
fprintf('rC_  =   [ %6.3f, %6.3f, %g ] (m)\n', rC_)
fprintf('phi2 = %6.3f (degrees) \n', phi2*180/pi)
fprintf('phi3 = %6.3f (degrees) \n', phi3*180/pi)
```

1.6 Velocity and Acceleration Analysis for Rigid Body

The motion of a rigid body (RB), with respect to a fixed reference frame, is defined by the position, velocity and acceleration of all points of the rigid body. A fixed orthogonal Cartesian reference frame $x_0 y_0 z_0$ with the constant unit vectors $\mathbf{I}_0, \mathbf{J}_0$, and $\mathbf{k}_0$, is shown in Fig. 1.12. The body fixed (mobile or rotating) orthogonal Cartesian

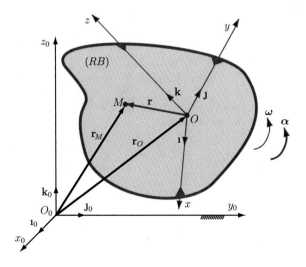

Fig. 1.12 Fixed orthogonal Cartesian reference frame with the unit vectors $[\mathbf{I}_0, \mathbf{J}_0, \mathbf{k}_0]$; body fixed (or rotating) reference frame with the unit vectors $[\mathbf{I}, \mathbf{J}, \mathbf{k}]$; point M is an arbitrary point on the rigid body, $M \in (RB)$

reference frame xyz with unit vectors $\imath, \jmath$ and $\mathbf{k}$ is attached to the moving rigid body. The unit vectors $\imath_0$, $\jmath_0$, and $\mathbf{k}_0$ of the primary reference frame are constant with respect to time and the unit vectors $\imath$, $\jmath$, and $\mathbf{k}$ are functions of time, t. The unit vectors $\imath, \jmath$, and $\mathbf{k}$ of the body-fixed frame of reference rotate with the body-fixed reference frame. The origin O is arbitrary. The position vector of any point M, $\forall M \in (RB)$, with respect to the fixed reference frame $x_0 y_0 z_0$ is $\mathbf{r}_M = \mathbf{r}_{O_0 M}$ and with respect to the rotating reference frame $Oxyz$ is denoted by $\mathbf{r} = \mathbf{r}_{OM}$. The position vector of the origin O of the rotating reference frame with respect to the fixed point O_0 is $\mathbf{r}_O = \mathbf{r}_{O_0 O}$. The position vector of M is

$$\mathbf{r}_M = \mathbf{r}_O + \mathbf{r} = \mathbf{r}_O + x\,\imath + y\,\jmath + z\,\mathbf{k}, \tag{1.9}$$

where x, y, and z represent the projections of the vector $\mathbf{r} = \mathbf{r}_{OM}$ on the rotating reference frame

$$\mathbf{r} = x\,\imath + y\,\jmath + z\,\mathbf{k}. \tag{1.10}$$

The distance between two points of the rigid body O and M is constant, $O \in (RB)$, and $M \in (RB)$. The components x, y and z of the vector $\mathbf{r}$ with respect to the rotating reference frame are constant. The unit vectors $\imath$, $\jmath$, and $\mathbf{k}$ are time-dependent vector functions. For the unit vectors of an orthogonal Cartesian reference frame, $\imath$, $\jmath$ and $\mathbf{k}$, there are the following relations

$$\imath \cdot \imath = 1, \quad \jmath \cdot \jmath = 1, \quad \mathbf{k} \cdot \mathbf{k} = 1, \quad \imath \cdot \jmath = 0, \quad \jmath \cdot \mathbf{k} = 0, \quad \mathbf{k} \cdot \imath = 0. \tag{1.11}$$

The velocity of an arbitrary point M of the rigid body with respect to the fixed reference frame $x_0 y_0 z_0$, is the derivative with respect to time of the position vector $\mathbf{r}_M$

$$\mathbf{v}_M = \frac{d\mathbf{r}_M}{dt} = \frac{d\mathbf{r}_{O_0 M}}{dt} = \frac{d\mathbf{r}_O}{dt} + \frac{d\mathbf{r}}{dt}$$
$$= \mathbf{v}_O + x\frac{d\imath}{dt} + y\frac{d\jmath}{dt} + z\frac{d\mathbf{k}}{dt} + \frac{dx}{dt}\imath + \frac{dy}{dt}\jmath + \frac{dz}{dt}\mathbf{k}, \tag{1.12}$$

where $\mathbf{v}_O = \dot{\mathbf{r}}_O$ is the velocity of the origin of the rotating reference frame $Oxyz$ with respect to the fixed reference frame. Because all the points in the rigid body maintain their relative position, their velocity relative to the rotating reference frame xyz is zero, i.e., $\dot{x} = \dot{y} = \dot{z} = 0$.

The velocity of point M is

$$\mathbf{v}_M = \mathbf{v}_O + x\frac{d\imath}{dt} + y\frac{d\jmath}{dt} + z\frac{d\mathbf{k}}{dt} = \mathbf{v}_O + x\,\dot{\imath} + y\,\dot{\jmath} + z\,\dot{\mathbf{k}}. \tag{1.13}$$

The derivative of the Eq. (1.11) with respect to time gives

$$\frac{d\imath}{dt} \cdot \imath = 0, \quad \frac{d\jmath}{dt} \cdot \jmath = 0, \quad \frac{d\mathbf{k}}{dt} \cdot \mathbf{k} = 0, \tag{1.14}$$

and

$$\frac{d\mathbf{\imath}}{dt} \cdot \mathbf{J} + \mathbf{\imath} \cdot \frac{d\mathbf{J}}{dt} = 0, \quad \frac{d\mathbf{J}}{dt} \cdot \mathbf{k} + \mathbf{J} \cdot \frac{d\mathbf{k}}{dt} = 0, \quad \frac{d\mathbf{k}}{dt} \cdot \mathbf{\imath} + \mathbf{k} \cdot \frac{d\mathbf{\imath}}{dt} = 0. \quad (1.15)$$

For Eq. (1.15) the following notation is employed

$$\frac{d\mathbf{\imath}}{dt} \cdot \mathbf{J} = -\mathbf{\imath} \cdot \frac{d\mathbf{J}}{dt} = \omega_z, \quad \frac{d\mathbf{J}}{dt} \cdot \mathbf{k} = -\mathbf{J} \cdot \frac{d\mathbf{k}}{dt} = \omega_x, \quad \frac{d\mathbf{k}}{dt} \cdot \mathbf{\imath} = -\mathbf{k} \cdot \frac{d\mathbf{\imath}}{dt} = \omega_y, \quad (1.16)$$

where ω_x, ω_y and ω_z may be considered as the projections of a vector $\boldsymbol{\omega}$

$$\boldsymbol{\omega} = \omega_x \mathbf{\imath} + \omega_y \mathbf{J} + \omega_z \mathbf{k}.$$

Using the *Poisson formulas*

$$\frac{d\mathbf{\imath}}{dt} = \boldsymbol{\omega} \times \mathbf{\imath}, \quad \frac{d\mathbf{J}}{dt} = \boldsymbol{\omega} \times \mathbf{J}, \quad \frac{d\mathbf{k}}{dt} = \boldsymbol{\omega} \times \mathbf{k}, \quad (1.17)$$

where $\boldsymbol{\omega}$ is the angular velocity vector of the rigid body, the velocity of the point M on the rigid body is

$$\mathbf{v}_M = \mathbf{v}_O + x\,\boldsymbol{\omega} \times \mathbf{\imath} + y\,\boldsymbol{\omega} \times \mathbf{J} + z\,\boldsymbol{\omega} \times \mathbf{k} = \mathbf{v}_O + \boldsymbol{\omega} \times (x\,\mathbf{\imath} + y\,\mathbf{J} + z\,\mathbf{k}),$$

or

$$\mathbf{v}_M = \mathbf{v}_O + \boldsymbol{\omega} \times \mathbf{r}. \quad (1.18)$$

Combining Eqs. (1.13) and (1.18) it results that

$$\frac{d\mathbf{r}}{dt} = \dot{\mathbf{r}} = \boldsymbol{\omega} \times \mathbf{r}. \quad (1.19)$$

The relation between the velocities $\mathbf{v}_M$ and $\mathbf{v}_O$ of two points M and O on the rigid body is

$$\mathbf{v}_M = \mathbf{v}_O + \boldsymbol{\omega} \times \mathbf{r}_{OM}, \quad (1.20)$$

or

$$\mathbf{v}_M = \mathbf{v}_O + \mathbf{v}_{MO}^{rel}, \quad (1.21)$$

where $\mathbf{v}_{MO}^{rel}$ is the relative velocity, for rotational motion, of M with respect to O and is given by

$$\mathbf{v}_{MO}^{rel} = \mathbf{v}_{MO} = \boldsymbol{\omega} \times \mathbf{r}_{OM}. \quad (1.22)$$

The relative velocity $\mathbf{v}_{MO}$ is perpendicular to the position vector $\mathbf{r}_{OM}$, $\mathbf{v}_{MO} \perp \mathbf{r}_{OM}$, and has the direction given by the angular velocity vector $\boldsymbol{\omega}$. The magnitude of the relative velocity is $|\mathbf{v}_{MO}| = v_{MO} = \omega r_{OM}$.

The acceleration of point $\forall M \in (RB)$ with respect to a fixed reference frame $O_0 x_0 y_0 z_0$, is the double derivative with respect to time of the position vector $\mathbf{r}_M$

$$\mathbf{a}_M = \ddot{\mathbf{r}}_M = \dot{\mathbf{v}}_M = \frac{d\mathbf{v}}{dt} = \frac{d}{dt}(\mathbf{v}_O + \boldsymbol{\omega} \times \mathbf{r}) = \frac{d\mathbf{v}_O}{dt} + \frac{d\boldsymbol{\omega}}{dt} \times \mathbf{r} + \boldsymbol{\omega} \times \frac{d\mathbf{r}}{dt}$$

$$= \dot{\mathbf{v}}_O + \dot{\boldsymbol{\omega}} \times \mathbf{r} + \boldsymbol{\omega} \times \dot{\mathbf{r}}. \tag{1.23}$$

The acceleration of the point O with respect to the fixed reference frame $O_0 x_0 y_0 z_0$ is

$$\mathbf{a}_O = \dot{\mathbf{v}}_O = \ddot{\mathbf{r}}_O. \tag{1.24}$$

The derivative of the vector $\boldsymbol{\omega}$ with respect to the time is the angular acceleration vector $\boldsymbol{\alpha}$ given by

$$\boldsymbol{\alpha} = \frac{d\boldsymbol{\omega}}{dt} = \frac{d\omega_x}{dt}\mathbf{1} + \frac{d\omega_y}{dt}\mathbf{J} + \frac{d\omega_z}{dt}\mathbf{k} + \omega_x \frac{d\mathbf{1}}{dt} + \omega_y \frac{d\mathbf{J}}{dt} + \omega_z \frac{d\mathbf{k}}{dt}$$

$$= \alpha_x \mathbf{1} + \alpha_y \mathbf{J} + \alpha_z \mathbf{k} + \omega_x \boldsymbol{\omega} \times \mathbf{1} + \omega_y \boldsymbol{\omega} \times \mathbf{J} + \omega_z \boldsymbol{\omega} \times \mathbf{k}$$

$$= \alpha_x \mathbf{1} + \alpha_y \mathbf{J} + \alpha_z \mathbf{k} + \boldsymbol{\omega} \times \boldsymbol{\omega} = \alpha_x \mathbf{1} + \alpha_y \mathbf{J} + \alpha_z \mathbf{k}, \tag{1.25}$$

where $\alpha_x = \dfrac{d\omega_x}{dt}, \alpha_y = \dfrac{d\omega_y}{dt}$, and $\alpha_z = \dfrac{d\omega_z}{dt}$. In the previous expression the Poisson formulas have been used. Using Eqs. (1.23)–(1.25) the acceleration of the point M is

$$\mathbf{a}_M = \mathbf{a}_O + \boldsymbol{\alpha} \times \mathbf{r} + \boldsymbol{\omega} \times (\boldsymbol{\omega} \times \mathbf{r}). \tag{1.26}$$

The relation between the accelerations $\mathbf{a}_M$ and $\mathbf{a}_O$ of two points M and O on the rigid body is

$$\mathbf{a}_M = \mathbf{a}_O + \boldsymbol{\alpha} \times \mathbf{r}_{OM} + \boldsymbol{\omega} \times (\boldsymbol{\omega} \times \mathbf{r}_{OM}). \tag{1.27}$$

In the case of planar motion, $\boldsymbol{\omega} = \omega \mathbf{k}$, $\mathbf{r}_{OM} = x_{OM}\mathbf{1} + y_{OM}\mathbf{J}$

$$\boldsymbol{\omega} \times (\boldsymbol{\omega} \times \mathbf{r}_{OM}) = -\omega^2 \mathbf{r}_{OM},$$

and Eq. (1.27) becomes

$$\mathbf{a}_M = \mathbf{a}_O + \boldsymbol{\alpha} \times \mathbf{r}_{OM} - \omega^2 \mathbf{r}_{OM}. \tag{1.28}$$

Equation (1.28) can be written as

$$\mathbf{a}_M = \mathbf{a}_O + \mathbf{a}_{MO}^{\text{rel}}, \tag{1.29}$$

where $\mathbf{a}_{MO}^{\text{rel}}$ is the relative acceleration, for rotational motion, of M with respect to O and is given by

$$\mathbf{a}_{MO}^{\text{rel}} = \mathbf{a}_{MO} = \mathbf{a}_{MO}^n + \mathbf{a}_{MO}^t. \tag{1.30}$$

The normal relative acceleration of M with respect to O is

$$\mathbf{a}_{MO}^n = \boldsymbol{\omega} \times (\boldsymbol{\omega} \times \mathbf{r}_{OM}), \qquad (1.31)$$

is parallel to the position vector $\mathbf{r}_{OM}$, $\mathbf{a}_{MO}^n \| \mathbf{r}_{OM}$, and has the direction towards the center of rotation, from M to O. The magnitude of the normal relative acceleration is

$$|\mathbf{a}_{MO}^n| = a_{MO}^n = \omega^2 r_{OM} = \frac{v_{MO}^2}{r_{OM}}.$$

The tangential relative acceleration of M with respect to O

$$\mathbf{a}_{MO}^t = \boldsymbol{\alpha} \times \mathbf{r}_{OM}, \qquad (1.32)$$

is perpendicular to the position vector $\mathbf{r}_{OM}$, $\mathbf{a}_{MO}^t \perp \mathbf{r}_{OM}$, and has the direction given by the angular acceleration $\boldsymbol{\alpha}$. The magnitude of the normal relative acceleration is

$$|\mathbf{a}_{MO}^t| = a_{MO}^t = \alpha \, r_{OM}.$$

Remarks:

1. The first derivatives of a vector $\mathbf{p}$ with respect to a scalar variable η in two reference frames RF_i and RF_j are related as follows

$$\frac{^{(j)}d\,\mathbf{p}}{d\eta} = \frac{^{(i)}d\,\mathbf{p}}{d\eta} + \boldsymbol{\omega}_{ij} \times \mathbf{p}, \qquad (1.33)$$

where $\boldsymbol{\omega}_{ij}$ is the rate of change of orientation of RF_i in RF_j with respect to η and $\dfrac{^{(j)}d\,\mathbf{p}}{d\eta}$ is the total derivative of $\mathbf{p}$ with respect to η in RF_j. The Cartesian unit vectors attached to the reference frame RF_i are $\mathbf{1}_i$, $\mathbf{J}_i$, $\mathbf{k}_i$. The vector $\mathbf{p}$ is written as $\mathbf{p} = p_x\, \mathbf{1}_i + p_y\, \mathbf{J}_i + p_z\, \mathbf{k}_i$ and the derivative of $\mathbf{p}$ with respect to η in RF_i is

$$\frac{^{(i)}d\,\mathbf{p}}{d\eta} = \frac{d\,p_x}{d\eta}\, \mathbf{1}_i + \frac{d\,p_y}{d\eta}\, \mathbf{J}_i + \frac{d\,p_z}{d\eta}\, \mathbf{k}_i. \qquad (1.34)$$

2. Let RF_i, $i = 1, 2, ..., n$ be n reference frames. The angular velocity of a rigid body r in the reference frame RF_n, can be expressed as

$$\boldsymbol{\omega}_{rn} = \boldsymbol{\omega}_{r1} + \boldsymbol{\omega}_{12} + \boldsymbol{\omega}_{23} + \cdots + \boldsymbol{\omega}_{n-1,n}. \qquad (1.35)$$

Next the motion of a point that moves relative to a rigid body is analyzed. Figure 1.13 shows a rigid body (RB) in motion relative to a primary reference frame with its origin at point O_0, $x_0 y_0 z_0$. The primary reference frame is a fixed reference frame or an Earth-fixed reference frame. A body-fixed reference frame moves with the rigid body. The body-fixed reference frame, xyz, has its origin at a point O of

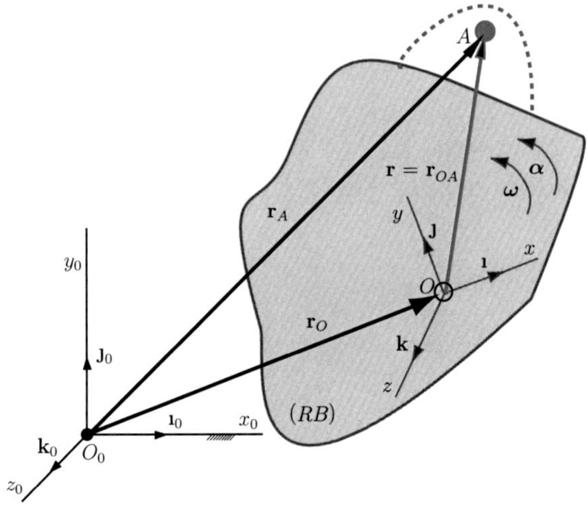

Fig. 1.13 The point A is not
assumed to be a point of the
moving rigid body,
$A \notin (RB)$

the rigid body, $O \in (RB)$. The unit vectors $\imath$, $\jmath$, and $\mathbf{k}$ of the body-fixed reference
rotate with the rigid body.

The position vector of point A (the point A is not assumed to be a point of the
rigid body, $A \notin (RB)$), relative to the origin O_0 of the primary reference frame is,
Fig. 1.13,

$$\mathbf{r}_A = \mathbf{r}_O + \mathbf{r},$$

where

$$\mathbf{r} = \mathbf{r}_{OA} = x\,\imath + y\,\jmath + z\,\mathbf{k}$$

is the position vector of A relative to the origin O, of the body-fixed reference frame,
and x, y, and z are the coordinates of A in terms of the body-fixed reference frame.
The velocity of the point A is the time derivative of the position vector $\mathbf{r}_A$

$$\mathbf{v}_A = \frac{d\mathbf{r}_O}{dt} + \frac{d\mathbf{r}}{dt} = \mathbf{v}_O + \mathbf{v}_{AO}^{\mathrm{rel}}$$

$$= \mathbf{v}_O + \frac{dx}{dt}\imath + x\frac{d\imath}{dt} + \frac{dy}{dt}\jmath + y\frac{d\jmath}{dt} + \frac{dz}{dt}\mathbf{k} + z\frac{d\mathbf{k}}{dt}.$$

With Poisson formulas, the total derivative of the position vector $\mathbf{r}$ is

$$\frac{d\mathbf{r}}{dt} = \dot{\mathbf{r}} = \dot{x}\,\imath + \dot{y}\,\jmath + \dot{z}\,\mathbf{k} + \boldsymbol{\omega} \times \mathbf{r}.$$

The velocity of A with respect to the body-fixed reference frame is a derivative in
the body-fixed reference frame

$$\mathbf{v}^{rel}_{A(xyz)} = \frac{^{(xyz)}d\,\mathbf{r}}{dt} = \frac{dx}{dt}\mathbf{\imath} + \frac{dy}{dt}\mathbf{J} + \frac{dz}{dt}\mathbf{k} = \dot{x}\mathbf{\imath} + \dot{y}\mathbf{J} + \dot{z}\mathbf{k}. \tag{1.36}$$

A general formula for the total derivative of a moving vector $\mathbf{r}$ may be written as

$$\frac{d\mathbf{r}}{dt} = \frac{^{(xyz)}d\,\mathbf{r}}{dt} + \boldsymbol{\omega} \times \mathbf{r}, \tag{1.37}$$

where $\dfrac{d\mathbf{r}}{dt} = \dfrac{^{(0)}d\,\mathbf{r}}{dt}$ is the derivative in the fixed (primary) reference frame (0) $(x_0 y_0 z_0)$, and $\dfrac{^{(xyz)}d\,\mathbf{r}}{dt}$ is the derivative in the rotating (mobile or body-fixed) reference frame (xyz).

The velocity of the point A relative to the fixed reference frame is

$$\mathbf{v}_A = \mathbf{v}_O + \mathbf{v}^{rel}_{A(xyz)} + \boldsymbol{\omega} \times \mathbf{r}. \tag{1.38}$$

The acceleration of the point A relative to the primary reference frame is obtained by taking the time derivative of Eq. (1.38)

$$\begin{aligned}\mathbf{a}_A &= \mathbf{a}_O + \mathbf{a}_{AO} \\ &= \mathbf{a}_O + \mathbf{a}^{rel}_{A(xyz)} + 2\boldsymbol{\omega} \times \mathbf{v}^{rel}_{A(xyz)} + \boldsymbol{\alpha} \times \mathbf{r} + \boldsymbol{\omega} \times (\boldsymbol{\omega} \times \mathbf{r}),\end{aligned} \tag{1.39}$$

where

$$\mathbf{a}^{rel}_{A(xyz)} = \frac{^{(xyz)}d^2\,\mathbf{r}}{dt^2} = \frac{d^2x}{dt^2}\mathbf{\imath} + \frac{d^2y}{dt^2}\mathbf{J} + \frac{d^2z}{dt^2}\mathbf{k}, \tag{1.40}$$

is the acceleration of A relative to the body-fixed reference frame or relative to the rigid body. The acceleration

$$\mathbf{a}^{cor}_{A(xyz)} = 2\boldsymbol{\omega} \times \mathbf{v}^{rel}_{A(xyz)}$$

is called the Coriolis relative acceleration. The direction of the Coriolis relative acceleration is obtained by rotating the linear relative velocity $\mathbf{v}^{rel}_{A(xyz)}$ through 90° in the direction of rotation given by $\boldsymbol{\omega}$.

In the case of planar motion, Eq. (1.39) becomes

$$\begin{aligned}\mathbf{a}_A &= \mathbf{a}_O + \mathbf{a}_{AO} \\ &= \mathbf{a}_O + \mathbf{a}^{rel}_{A(xyz)} + 2\boldsymbol{\omega} \times \mathbf{v}^{rel}_{A(xyz)} + \boldsymbol{\alpha} \times \mathbf{r} - \boldsymbol{\omega}^2 \mathbf{r}.\end{aligned} \tag{1.41}$$

The velocity $\mathbf{v}_A$ and the acceleration $\mathbf{a}_A$ of a point A are relative to the primary reference frame (absolute velocity and acceleration). The relative velocity $\mathbf{v}^{rel}_{A(xyz)}$ and relative acceleration $\mathbf{a}^{rel}_{A(xyz)}$ are the velocity and acceleration of point A relative to the body-fixed reference frame, i.e., they are the velocity and acceleration measured by an observer rigidly moving with the rigid body. If A is a point of the rigid body, $A \in (RB)$, $\mathbf{v}_{A(xyz)}^{\mathrm{rel}} = \mathbf{0}$ and $\mathbf{a}_{A(xyz)}^{\mathrm{rel}} = \mathbf{0}$.

The velocity and acceleration of an arbitrary point A relative to a point O of a rigid body, in terms of the body-fixed reference frame, are given by Eqs. (1.38) and (1.39)

$$\mathbf{v}_A = \mathbf{v}_O + \mathbf{v}_{AO}^{\mathrm{rel}} + \boldsymbol{\omega} \times \mathbf{r}_{OA}, \tag{1.42}$$

$$\mathbf{a}_A = \mathbf{a}_O + \mathbf{a}_{AO}^{\mathrm{rel}} + 2\,\boldsymbol{\omega} \times \mathbf{v}_{AO}^{\mathrm{rel}} + \boldsymbol{\alpha} \times \mathbf{r}_{OA} + \boldsymbol{\omega} \times (\boldsymbol{\omega} \times \mathbf{r}_{OA}). \tag{1.43}$$

These results apply to any reference frame having a moving origin O and rotating with angular velocity $\boldsymbol{\omega}$ and angular acceleration $\boldsymbol{\alpha}$ relative to a primary reference frame. The terms $\mathbf{v}_{AO}^{\mathrm{rel}}$ and $\mathbf{a}_{AO}^{\mathrm{rel}}$ are the velocity and acceleration of A with respect to the rotating reference frame, i.e., they are the velocity and acceleration measured by an observer moving with the rotating reference frame.

1.7 Planar Dynamics Analysis

Consider an arbitrary rigid body in motion with the total mass m. The position of the mass center of the rigid body is $\mathbf{r}_C$ and $\mathbf{a}_C$ is the acceleration of the mass center C. The sum of the external forces, $\mathbf{F}$, acting on the system equals the product of the mass and the acceleration of the mass center

$$m\,\mathbf{a}_C = \mathbf{F}. \tag{1.44}$$

Equation (1.44) is Newton's second law for a rigid body and is applicable to planar and three-dimensional motions. Resolving the sum of the external forces into Cartesian rectangular components

$$\mathbf{F} = F_x\,\mathbf{\imath} + F_y\,\mathbf{\jmath} + F_z\,\mathbf{k},$$

and the position vector of the mass center

$$\mathbf{r}_C = x_C(t)\,\mathbf{\imath} + y_C(t)\,\mathbf{\jmath} + z_C(t)\,\mathbf{k},$$

Newton's second law for the rigid body is

$$m\,\ddot{\mathbf{r}}_C = \mathbf{F}, \tag{1.45}$$

or

$$m\,\ddot{x}_C = F_x, \quad m\,\ddot{y}_C = F_y, \quad m\,\ddot{z}_C = F_z. \tag{1.46}$$

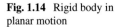

Fig. 1.14 Rigid body in
planar motion

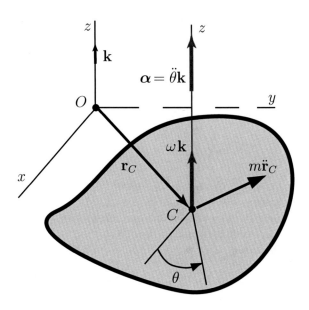

Figure 1.14 shows a rigid body moving in the (x, y) plane with the origin at O.
The mass center C of the rigid body is in the plane of the motion. The axis Oz
is perpendicular to the plane of motion of the rigid body, where O is the origin.
The axis Cz is perpendicular to the plane of motion through the mass center C. The
angular velocity vector of the rigid body is $\boldsymbol{\omega} = \omega\,\mathbf{k}$ and the angular acceleration is
$\boldsymbol{\alpha} = \dot{\boldsymbol{\omega}} = \ddot{\theta}\,\mathbf{k}$, where the angle θ is the angular position of the rigid body about a
fixed axis. The mass moment of inertia of the rigid body about the z-axis through C
is also the polar mass moment of inertia of the rigid body about C, $I_{Cz} = I_C$.

The rotational equation (Euler's equation) of motion for the rigid body is

$$I_C\,\boldsymbol{\alpha} = \sum \mathbf{M}_C, \tag{1.47}$$

where $\mathbf{M}_C$ is the sum of the moments about C due to external forces and couples.

Consider the case when the rigid body rotates about a fixed point P. The mass
moment of inertia of the rigid body about the z-axis through P is (parallel-axis
theorem)

$$I_{Pz} = I_{Cz} + m\,(CP)^2.$$

For the rigid body rotation about a fixed point P, Euler's equation is

$$I_{Pz}\boldsymbol{\alpha} = \sum \mathbf{M}_P, \tag{1.48}$$

where $\mathbf{M}_P$ is the sum of the moments about the fixed point P due to external forces
and couples.

The Newton-Euler general equations of motion for a rigid body in plane motion are

$$\mathbf{F} = m\,\mathbf{a}_C \quad \text{or} \quad \mathbf{F} = m\,\ddot{\mathbf{r}}_C,$$
$$\sum \mathbf{M}_C = I_{Cz}\,\boldsymbol{\alpha}, \tag{1.49}$$

or

$$m\,\ddot{x}_C = \sum F_x, \quad m\,\ddot{y}_C = \sum F_y, \quad I_{Cz}\,\ddot{\theta} = \sum M_C. \tag{1.50}$$

Equations (1.49) and (1.50), can be applied in two ways: the forces and moments are known and the equations are solved for the motion of the rigid body (direct dynamics); the motion of the rigid body is known and the equations are solved for the forces and moments (inverse dynamics).

D'Alembert's Principle. Newton's second law can be written as

$$\mathbf{F} + (-m\,\mathbf{a}_C) = \mathbf{0}, \quad \text{or} \quad \mathbf{F} + \mathbf{F}_{\text{in}} = \mathbf{0},$$

where the term $\mathbf{F}_{\text{in}} = -m\,\mathbf{a}_C$ is the inertia force.

Euler's equation moment about C is

$$\sum M_C + (-I_{Cz}\,\ddot{\theta}) = \mathbf{0}, \quad \text{or} \quad \sum M_C + \mathbf{M}_{\text{in}} = \mathbf{0}.$$

The term $\mathbf{M}_{\text{in}} = -I_{Cz}\,\boldsymbol{\alpha}$ is the inertia moment.

The equations of motion for a rigid body are similar to the equations for static equilibrium: the sum of the forces equals zero and the sum of the moments about any point equals zero when the inertia force and moment are taken into account. This is called D'Alembert's principle. With d'Alembert's principle the moment summation is about any arbitrary point Q

$$\sum \mathbf{M}_Q + \mathbf{M}_{\text{in}} + \mathbf{r}_{QC} \times \mathbf{F}_{\text{in}} = \mathbf{0},$$

where $\sum \mathbf{M}_Q$ is the sum of all external moments about Q, $\mathbf{M}_{\text{in}}$ is the inertia moment, $\mathbf{F}_{\text{in}}$ is the inertia force, and $\mathbf{r}_{QC}$ is a vector from Q to C.

In this way the dynamic analysis can be transformed to a static force and moment equilibrium problem. The inertia force and the moment are considered as external force and moment.

Lagrange's equations of motion. For a system of particles $\{S\}$ with n degrees of freedom, the position vector $\mathbf{r}_i$ of a particle $P_i \in \{S\}$ is

$$\mathbf{r}_i = \mathbf{r}_i(q_1, \dots, q_n, t), \tag{1.51}$$

where $q_1, ..., q_n$ are the generalized coordinates and t is the time. The D'Alembert's principle for the system is

$$\mathbf{F}_i - m_i \ddot{\mathbf{r}}_i = 0, \quad i = 1, ..., n, \tag{1.52}$$

where m_i is the mass of P_i, $\ddot{\mathbf{r}}_i$ is the acceleration of P_i, and $\mathbf{F}_i$ is the resultant of the net forces acting on P_i. The velocity of the particle P_i is

$$\mathbf{v}_i = \sum_{r=1}^{n} \frac{\partial \mathbf{r}_i}{\partial q_r} \frac{\partial q_r}{\partial t} + \frac{\partial \mathbf{r}_i}{\partial t} = \sum_{r=1}^{n} \frac{\partial \mathbf{r}_i}{\partial q_r} \dot{q}_r + \frac{\partial \mathbf{r}_i}{\partial t}, \tag{1.53}$$

and

$$\frac{\partial \mathbf{v}_i}{\partial \dot{q}_r} = \frac{\partial \mathbf{r}_i}{\partial q_r}. \tag{1.54}$$

The D'Alembert's principle can be written as

$$\sum_{i=1}^{n} (\mathbf{F}_i - m_i \ddot{\mathbf{r}}_i) \cdot \frac{\partial \mathbf{r}_i}{\partial q_r} = 0, \quad r = 1, ..., n. \tag{1.55}$$

The generalized active force, Q_r, is defined as

$$Q_r = \sum_{i=1}^{n} \frac{\partial \mathbf{r}_i}{\partial q_r} \cdot \mathbf{F}_i = \sum_{i=1}^{n} \frac{\partial \mathbf{v}_i}{\partial \dot{q}_r} \cdot \mathbf{F}_i, \quad r = 1, ..., n. \tag{1.56}$$

The generalized inertia force, $K_{in\,r}$ are defined as [30, 31]

$$K_{in\,r} = \sum_{i=1}^{n} \frac{\partial \mathbf{r}_i}{\partial q_r} \cdot (-m_i \ddot{\mathbf{r}}_i) = \sum_{i=1}^{n} \frac{\partial \mathbf{v}_i}{\partial \dot{q}_r} \cdot (-m_i \ddot{\mathbf{r}}_i), \quad r = 1, ..., n. \tag{1.57}$$

Kane's dynamical equations for the system are [30, 31]

$$Q_r + K_{in\,r} = 0, \quad r = 1, ..., n. \tag{1.58}$$

The kinetic energy of the system $\{S\}$ is

$$T = \frac{1}{2} \sum_{i=1}^{n} m_i \mathbf{v}_i \cdot \mathbf{v}_i. \tag{1.59}$$

The relation between the generalized inertia forces and the kinetic energy is [30, 31]

$$K_{in\,r} = -\left[\frac{d}{dt} \left(\frac{\partial T}{\partial \dot{q}_r} \right) - \frac{\partial T}{\partial q_r} \right], \quad r = 1, ..., n.$$

The Lagrange's equations of motion of the first kind are

$$\frac{d}{dt}\left(\frac{\partial T}{\partial \dot{q}_r}\right) - \frac{\partial T}{\partial q_r} = Q_r, \quad r = 1, ..., n. \qquad (1.60)$$

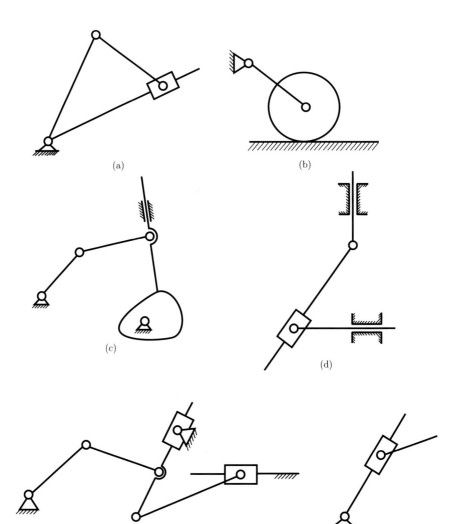

(a)

(b)

(c)

(d)

(e)

(f)

Fig. 1.15 Problem 1.1

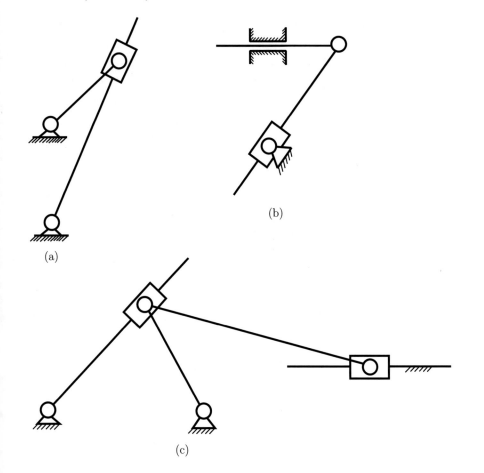

(b)

(a)

(c)

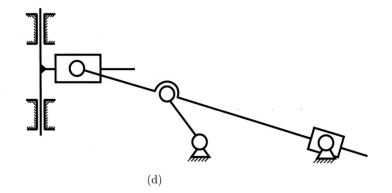

(d)

Fig. 1.16 Problem 1.2

1.8 Problems

1.1 Find the numbers of degrees of freedom for the mechanical systems in Fig. 1.15.
1.2 Find the numbers of degrees of freedom, the type of dyads, the contour diagrams, for the mechanical systems in Fig. 1.16.

References

1. E.A. Avallone, T. Baumeister, A. Sadegh, *Marks' Standard Handbook for Mechanical Engineers*, 11th edn. (McGraw-Hill Education, New York, 2007)
2. I.I. Artobolevski, *Mechanisms in Modern Engineering Design* (MIR, Moscow, 1977)
3. M. Atanasiu, *Mechanics (Mecanica)* (EDP, Bucharest, 1973)
4. H. Baruh, *Analytical Dynamics* (WCB/McGraw-Hill, Boston, 1999)
5. A. Bedford, W. Fowler, *Dynamics* (Addison Wesley, Menlo Park, CA, 1999)
6. A. Bedford, W. Fowler, *Statics* (Addison Wesley, Menlo Park, CA, 1999)
7. F.P. Beer et al., *Vector Mechanics for Engineers: Statics and Dynamics* (McGraw-Hill, New York, NY, 2016)
8. M. Buculei, D. Bagnaru, G. Nanu, D.B. Marghitu, *Computing Methods in the Analysis of the Mechanisms with Bars* (Scrisul Romanesc, Craiova, 1986)
9. M.I. Buculei, *Mechanisms* (University of Craiova Press, Craiova, Romania, 1976)
10. D. Bolcu, S. Rizescu, *Mecanica* (EDP, Bucharest, 2001)
11. J. Billingsley, *Essential of Dynamics and Vibration* (Springer, 2018)
12. R. Budynas, K.J. Nisbett, *Shigley's Mechanical Engineering Design*, 9th edn. (McGraw-Hill, New York, 2013)
13. J.A. Collins, H.R. Busby, and G.H. Staab, *Mechanical Design of Machine Elements and Machines* (2nd ed., Wiley, 2009)
14. M. Crespo da Silva, *Intermediate Dynamics for Engineers* (McGraw-Hill, New York, 2004)
15. A.G. Erdman, G.N. Sandor, *Mechanisms Design* (Prentice-Hall, Upper Saddle River, NJ, 1984)
16. M. Dupac, D.B. Marghitu, *Engineering Applications: Analytical and Numerical Calculation with MATLAB* (Wiley, Hoboken, NJ, 2021)
17. A. Ertas, J.C. Jones, *The Engineering Design Process* (Wiley, New York, 1996)
18. F. Freudenstein, An application of Boolean Algebra to the motion of epicyclic drivers. J. Eng. Ind. 176–182 (1971)
19. J.H. Ginsberg, *Advanced Engineering Dynamics* (Cambridge University Press, Cambridge, 1995)
20. H. Goldstein, *Classical Mechanics* (Addison-Wesley, Redwood City, CA, 1989)
21. D.T. Greenwood, *Principles of Dynamics* (Prentice-Hall, Englewood Cliffs, NJ, 1998)
22. A.S. Hall, A.R. Holowenko, H.G. Laughlin, *Schaum's Outline of Machine Design* (McGraw-Hill, New York, 2013)
23. B.G. Hamrock, B. Jacobson, S.R. Schmid, *Fundamentals of Machine Elements* (McGraw-Hill, New York, 1999)
24. R.C. Hibbeler, *Engineering Mechanics: Dynamics* (Prentice Hall, 2010)
25. T.E. Honein, O.M. O'Reilly, On the Gibbs–Appell equations for the dynamics of rigid bodies. J. Appl. Mech. 88/074501-1 (2021)
26. R.C. Juvinall, K.M. Marshek, *Fundamentals of Machine Component Design*, 5th edn. (Wiley, New York, 2010)
27. T.R. Kane, *Analytical Elements of Mechanics*, vol. 1 (Academic Press, New York, 1959)
28. T.R. Kane, *Analytical Elements of Mechanics*, vol. 2 (Academic Press, New York, 1961)

29. T.R. Kane, D.A. Levinson, The use of Kane's dynamical equations in robotics. MIT Int. J. Robot. Res. **3**, 3–21 (1983)
30. T.R. Kane, P.W. Likins, D.A. Levinson, *Spacecraft Dynamics* (McGraw-Hill, New York, 1983)
31. T.R. Kane, D.A. Levinson, *Dynamics* (McGraw-Hill, New York, 1985)
32. K. Lingaiah, *Machine Design Databook*, 2nd edn. (McGraw-Hill Education, New York, 2003)
33. N.I. Manolescu, F. Kovacs, A. Oranescu, *The Theory of Mechanisms and Machines* (EDP, Bucharest, 1972)
34. D.B. Marghitu, *Mechanical Engineer's Handbook* (Academic Press, San Diego, CA, 2001)
35. D.B. Marghitu, M.J. Crocker, *Analytical Elements of Mechanisms* (Cambridge University Press, Cambridge, 2001)
36. D.B. Marghitu, *Kinematic Chains and Machine Component Design* (Elsevier, Amsterdam, 2005)
37. D.B. Marghitu, *Mechanisms and Robots Analysis with MATLAB* (Springer, New York, NY, 2009)
38. D.B. Marghitu, M. Dupac, *Advanced Dynamics: Analytical and Numerical Calculations with MATLAB* (Springer, New York, NY, 2012)
39. D.B. Marghitu, M. Dupac, H.M. Nels, *Statics with MATLAB* (Springer, New York, NY, 2013)
40. D.B. Marghitu, D. Cojocaru, *Advances in Robot Design and Intelligent Control* (Springer International Publishing, Cham, Switzerland, 2016), pp. 317–325
41. D.J. McGill, W.W. King, *Engineering Mechanics: Statics and an Introduction to Dynamics* (PWS Publishing Company, Boston, 1995)
42. J.L. Meriam, L.G. Kraige, *Engineering Mechanics: Dynamics* (Wiley, New York, 2007)
43. C.R. Mischke, Prediction of stochastic endurance strength. Trans. ASME J. Vib. Acoust. Stress Reliab. Des. **109**(1), 113–122 (1987)
44. L. Meirovitch, *Methods of Analytical Dynamics* (Dover, 2003)
45. R.L. Mott, *Machine Elements in Mechanical Design* (Prentice Hall, Upper Saddle River, NJ, 1999)
46. W.A. Nash, *Strength of Materials* (Schaum's Outline Series, McGraw-Hill, New York, 1972)
47. R.L. Norton, *Machine Design* (Prentice-Hall, Upper Saddle River, NJ, 1996)
48. R.L. Norton, *Design of Machinery* (McGraw-Hill, New York, 1999)
49. O.M. O'Reilly, *Intermediate Dynamics for Engineers Newton-Euler and Lagrangian Mechanics* (Cambridge University Press, UK, 2020)
50. O.M. O'Reilly, *Engineering Dynamics: A Primer* (Springer, NY, 2010)
51. W.C. Orthwein, *Machine Component Design* (West Publishing Company, St. Paul, 1990)
52. L.A. Pars, *A Treatise on Analytical Dynamics* (Wiley, New York, 1965)
53. F. Reuleaux, *The Kinematics of Machinery* (Dover, New York, 1963)
54. D. Planchard, M. Planchard, *SolidWorks 2013 Tutorial with Video Instruction* (SDC Publications, 2013)
55. I. Popescu, *Mechanisms* (University of Craiova Press, Craiova, Romania, 1990)
56. I. Popescu, C. Ungureanu, *Structural Synthesis and Kinematics of Mechanisms with Bars* (Universitaria Press, Craiova, Romania, 2000)
57. I. Popescu, L. Luca, M. Cherciu, D.B. Marghitu, *Mechanisms for Generating Mathematical Curves* (Springer Nature, Switzerland, 2020)
58. C.A. Rubin, *The Student Edition of Working Model* (Addison-Wesley Publishing Company, Reading, MA, 1995)
59. J. Ragan, D.B. Marghitu, Impact of a Kinematic link with MATLAB and SolidWorks. Applied Mechanics and Materials **430**, 170–177 (2013)
60. J. Ragan, D.B. Marghitu, MATLAB dynamics of a free link with elastic impact, in *International Conference on Mechanical Engineering, ICOME 2013*, 16–17 May 2013 (Craiova, Romania, 2013)
61. J.C. Samin, P. Fisette, *Symbolic Modeling of Multibody Systems* (Kluwer, 2003)
62. A.A. Shabana, *Computational Dynamics* (Wiley, New York, 2010)
63. I.H. Shames, *Engineering Mechanics—Statics and Dynamics* (Prentice-Hall, Upper Saddle River, NJ, 1997)

64. J.E. Shigley, C.R. Mischke, *Mechanical Engineering Design* (McGraw-Hill, New York, 1989)
65. J.E. Shigley, C.R. Mischke, R.G. Budynas, *Mechanical Engineering Design*, 7th edn. (McGraw-Hill, New York, 2004)
66. J.E. Shigley, J.J. Uicker, *Theory of Machines and Mechanisms* (McGraw-Hill, New York, 1995)
67. D. Smith, *Engineering Computation with MATLAB* (Pearson Education, Upper Saddle River, NJ, 2008)
68. R.W. Soutas-Little, D.J. Inman, *Engineering Mechanics: Statics and Dynamics* (Prentice-Hall, Upper Saddle River, NJ, 1999)
69. J. Sticklen, M.T. Eskil, *An Introduction to Technical Problem Solving with MATLAB* (Great Lakes Press, Wildwood, MO, 2006)
70. A.C. Ugural, *Mechanical Design* (McGraw-Hill, New York, 2004)
71. R. Voinea, D. Voiculescu, V. Ceausu, *Mechanics (Mecanica)* (EDP, Bucharest, 1983)
72. K.J. Waldron, G.L. Kinzel, *Kinematics, Dynamics, and Design of Machinery* (Wiley, New York, 1999)
73. J.H. Williams Jr., *Fundamentals of Applied Dynamics* (Wiley, New York, 1996)
74. C.E. Wilson, J.P. Sadler, *Kinematics and Dynamics of Machinery* (Harper Collins College Publishers, New York, 1991)
75. H.B. Wilson, L.H. Turcotte, D. Halpern, *Advanced Mathematics and Mechanics Applications Using MATLAB* (Chapman & Hall/CRC, 2003)
76. S. Wolfram, *Mathematica* (Wolfram Media/Cambridge University Press, Cambridge, 1999)
77. National Council of Examiners for Engineering and Surveying (NCEES), *Fundamentals of Engineering. Supplied-Reference Handbook* (Clemson, SC, 2001)
78. * * *, *The Theory of Mechanisms and Machines* (Teoria mehanizmov i masin) (Vassaia scola, Minsk, Russia, 1970)
79. eCourses—University of Oklahoma. http://ecourses.ou.edu/home.htm
80. https://www.mathworks.com
81. http://www.eng.auburn.edu/~marghitu/
82. https://www.solidworks.com
83. https://www.wolfram.com

Chapter 2
Classical Analysis of a Mechanism with One Dyad

Abstract The planar motion of a mechanism with three moving links is analyzed. Symbolical and numerical MATLAB are used for the kinematics and dynamics of the system. The classical vectorial equations for velocity and acceleration of the rigid body are used. The joint reaction forces and the moment applied to the driver link are calculated for a given position with Newton-Euler equations for inverse dynamics.

The planar R-RTR mechanism considered is shown in Fig. 2.1. The driver link is the rigid link 1 (the link AB). The dyad RTR (B_R, B_T, C_R) is composed of the slider 2 and the rocker 3. The following numerical data are given: $AB = 0.10$ m, $AC = 0.05$ m, and $CD = 0.15$ m. The angle of the driver link 1 with the horizontal axis is $\phi = 30°$. The constant angular speed of the driver link 1 is -50 rpm.

Given an external moment $\mathbf{M}_e = -100 \operatorname{sign}(\omega_3)\,\mathbf{k}$ Nm applied on the link 3, calculate the motor moment $\mathbf{M}_m$ required for the dynamic equilibrium of the mechanism. All three links are rectangular prisms with the depth $d = 0.001$ m and the mass density $\rho = 8000$ kg/m^3. The height of the links 1 and 3 is $h = 0.01$ m. The slider 2 has the height $h_S = 0.02$ m, and the width $w_S = 0.04$ m. The center of mass location of the links $i = 1, 2, 3$ are designated by $C_i(x_{C_i}, y_{C_i}, 0)$. The gravitational acceleration is $g = 9.807$ m/s^2.

2.1 Position Analysis

A Cartesian reference frame xy is selected. The joint A is in the origin of the reference frame, that is, $A \equiv O$,

$$x_A = 0 \quad \text{and} \quad y_A = 0.$$

The coordinates of the joint C are

$$x_C = AC = 0.05 \quad \text{and} \quad y_C = 0 \text{ m}.$$

© The Author(s), under exclusive license to Springer Nature Switzerland AG 2022
D. B. Marghitu et al., *Mechanical Simulation with MATLAB®*,
Springer Tracts in Mechanical Engineering,
https://doi.org/10.1007/978-3-030-88102-3_2

Fig. 2.1 R-RTR mechanism

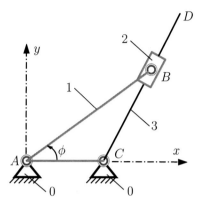

Position of joint B

The unknowns are the coordinates of the joint B, x_B and y_B. Because the joint A is fixed and the angle ϕ is known, the coordinates of the joint B are computed from the following expressions

$$x_B = AB \cos \phi = 0.10 \cos 30° = 0.087 \text{ m},$$
$$y_B = AB \sin \phi = 0.10 \sin 30° = 0.050 \text{ m}.$$

The MATLAB commands for joints A, C, and B are:

```
AB = 0.10 ;    % (m)
AC = 0.05 ;    % (m)
CD = 0.15 ;    % (m)
phi = pi/6;    % (rad)
xA = 0; yA = 0;
rA_ = [xA yA 0];
xC = AC ; yC = 0;
rC_ = [xC yC 0];
xB = AB*cos(phi);
yB = AB*sin(phi);
rB_ = [xB yB 0];
```

Angle ϕ_2

The angle of link 2 (or link 3) with the horizontal axis is calculated from the slope of the straight line BC:

$$\phi_2 = \phi_3 = \arctan \frac{y_B - y_C}{x_B - x_C} = \arctan \frac{0.050}{0.087 - 0.050} = 0.939 \text{ rad} = 53.794°.$$

Position of joint D
The unknowns are the coordinates of the joint D, x_D and y_D

$$x_D = x_C + CD \cos \phi_3 = 0.050 + 0.15 \cos 53.794° = 0.139 \text{ m},$$
$$y_D = y_C + CD \sin \phi_3 = 0.15 \sin 53.794° = 0.121 \text{ m}.$$

The MATLAB commands for the position of D are:

```
phi2 = atan((yB-yC)/(xB-xC));
phi3 = phi2;
CB = norm([xB-xC, yB-yC]);
ux = (xB - xC)/CB;
uy = (yB - yC)/CB;
xD = xC + CD*ux;
yD = yC + CD*uy;
rD_ = [xD yD 0];
```

The components of the unit vector of the vector $\overrightarrow{CD}$ are ux and uy. The results are printed in MATLAB with:

```
fprintf('Results \n\n')
fprintf('phi = phi1 = %g (degrees) \n',phi*180/pi)
fprintf('rA_ = [%6.3f,%6.3f,%g] (m)\n',rA_)
fprintf('rC_ = [%6.3f,%6.3f,%g] (m)\n',rC_)
fprintf('rB_ = [%6.3f,%6.3f,%g] (m)\n',rB_)
fprintf('phi2 = phi3 = %6.3f (degrees) \n',phi2*180/pi)
fprintf('rD_ = [%6.3f,%6.3f,%g] (m)\n',rD_)
```

and the numerical output is

```
phi = phi1 = 30 (degrees)
rA_ = [ 0.000, 0.000,0] (m)
rC_ = [ 0.050, 0.000,0] (m)
rB_ = [ 0.087, 0.050,0] (m)
phi2 = phi3 = 53.794 (degrees)
rD_ = [ 0.139, 0.121,0] (m)
```

The MATLAB representation for R-RTR mechanism with $\phi = \pi/6$ is shown in Fig. 2.2. The graphic is obtained with the statements:

```
% Graphic of the mechanism
axis manual
axis equal
hold on
grid on
```

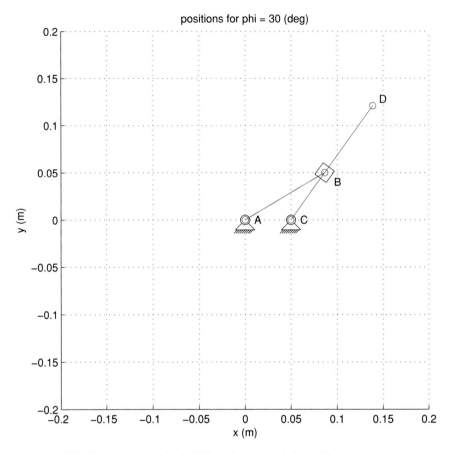

Fig. 2.2 MATLAB representation for R-RTR mechanism with $\phi = \pi/6$

```
ax = 0.2;
axis([-ax ax -ax ax])
xlabel('x (m)'), ylabel('y (m)')
title('positions for phi = 30 (deg)')
textAx = xA+0.01;
textAy = yA;
textBx = xB+0.01;
textBy = yB-0.01;
textCx = xC+0.01;
textCy = yC;
textDx = xD+0.007;
textDy = yD+0.007;
text_size = 10;
slider_scale_factor = 0.015;
```

```
pivot_scale_factor = 0.005;
% plotting joint A
plot_pin0(xA,yA,pivot_scale_factor,0)
% labelling joint A
text(textAx, textAy,'A','FontSize',text_size)
% plotting link 1
plot([xA xB], [yA yB], '-o')
% labelling joint B
text(textBx, textBy,'B','FontSize',text_size)
% plotting link 2
plot_slider( xB, yB, slider_scale_factor, phi2)
% plotting joint C
plot_pin0(xC,yC,pivot_scale_factor,0)
% labelling joint C
text( textCx, textCy,'C','FontSize',text_size)
% plotting link 3
plot([xC xD], [yC yD], '-o')
% labelling joint D
text(textDx, textDy,'D','FontSize',text_size)
```

For the graphical representation the following external graphical functions have been used: plot_pin0.m to plot the pin joints at A and C and plot_slider.m to plot the slider joint at B between links 2 and 3 [81].

The position simulation for a complete rotation of the driver link 1 ($\phi \in [0, 360°]$) is obtained with the MATLAB commands:

```
step=pi/6;
for phi=pi/6:step:2*pi+pi/6
xB = AB*cos(phi); yB = AB*sin(phi);
rB_ = [xB yB 0];
CB = norm([xB-xC, yB-yC]);
cx = (xB - xC)/CB; cy = (yB - yC)/CB;
xD = xC + CD*cx; yD = yC + CD*cy;
% Graphic of the mechanism
text_size = 10;
slider_scale_factor = 0.015;
pin_scale_factor = 0.005;
plot_pin0(xA,yA,pin_scale_factor,0)
plot_pin0(xC,yC,pin_scale_factor,0)
plot_slider( xB, yB, slider_scale_factor, phi2);
hold on
plot(...
[xA,xB],[yA,yB],'r-o',...
[xC,xD],[yC,yD],'b-o');
plot_slider( xB, yB, slider_scale_factor, phi2);
hold off
ax = 0.2;
axis([-ax ax -ax ax])
grid on
```

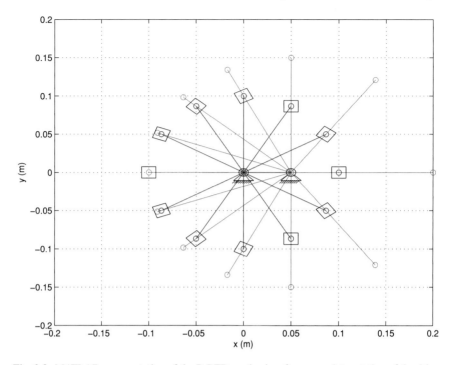

Fig. 2.3 MATLAB representation of the R-RTR mechanism for a complete rotation of the driver link

```
xlabel('x (m)'), ylabel('y (m)')
pause(0.1)
end
```

The graphical representation of the mechanism for a complete rotation of the link 1 is shown in Fig. 2.3. The position of a point E of the mechanism is calculated with:

```
b = 0.08;
xE = xB - b*ux;
yE = yB - b*uy;
```

The trajectory of the point E for a complete rotation of the driver link 1 is depicted in Fig. 2.4 and it is obtained with the commands:

```
step=pi/18;
for phi=pi/6:step:2*pi+pi/6
xB = AB*cos(phi); yB = AB*sin(phi);
rB_ = [xB yB 0];
CB = norm([xB-xC, yB-yC]);
ux = (xB - xC)/CB; uy = (yB - yC)/CB;
```

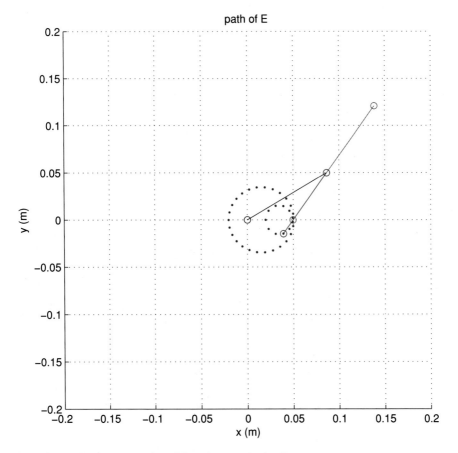

Fig. 2.4 MATLAB representation of the trajectory of point *E*

```
xD = xC + CD*ux; yD = yC + CD*uy;
%
b = 0.08;
xE = xB - b*ux;
yE = yB - b*uy;
%
axis manual
axis equal
hold on
grid on
ax = 0.2;
axis([-ax ax -ax ax])
xlabel('x (m)'), ylabel('y (m)')
title('path of E')
```

```
pM=plot(...
[xA,xB],[yA,yB],'k-o',...
[xC,xD],[yC,yD],'b-o',...
[xB,xE],[yB,yE],'r-o');
plot(xE,yE,'.')
pause(1)
delete(pM)
end
pM=plot(...
[xA,xB],[yA,yB],'k-o',...
[xC,xD],[yC,yD],'b-o',...
[xB,xE],[yB,yE],'r-o');
```

2.2 Velocity and Acceleration Analysis

The velocity of the point B_1 on the link 1 is

$$\mathbf{v}_{B_1} = \mathbf{v}_A + \boldsymbol{\omega}_1 \times \mathbf{r}_{AB} = \boldsymbol{\omega}_1 \times \mathbf{r}_B,$$

where $\mathbf{v}_A \equiv \mathbf{0}$ is the velocity of the origin $A \equiv O$.
The angular velocity of link 1 is

$$\boldsymbol{\omega}_1 = \omega_1 \, \mathbf{k} = \frac{\pi n}{30} \, \mathbf{k} = \frac{\pi(-50)}{30} \, \mathbf{k} = -5.236 \, \mathbf{k} \ \text{rad/s}.$$

The position vector of point B is

$$\mathbf{r}_{AB} = \mathbf{r}_B - \mathbf{r}_A = \mathbf{r}_B = x_B \, \boldsymbol{\imath} + y_B \, \mathbf{J} + z_B \, \mathbf{k} = 0.087 \, \boldsymbol{\imath} + 0.050 \, \mathbf{J} \ \text{m}.$$

The velocity of point B_2 on the link 2 is $\mathbf{v}_{B_2} = \mathbf{v}_{B_1}$ because between the links 1 and 2 there is a rotational joint. The velocity of $B_1 = B_2$ is

$$\mathbf{v}_{B_1} = \mathbf{v}_{B_2} = \begin{vmatrix} \boldsymbol{\imath} & \mathbf{J} & \mathbf{k} \\ 0 & 0 & \omega \\ x_B & y_B & 0 \end{vmatrix} = \begin{vmatrix} \boldsymbol{\imath} & \mathbf{J} & \mathbf{k} \\ 0 & 0 & -5.236 \\ 0.087 & 0.050 & 0 \end{vmatrix} = 0.262 \, \boldsymbol{\imath} - 0.453 \, \mathbf{J} \ \text{m/s}.$$

The acceleration of the point $B_1 = B_2$ is

$$\mathbf{a}_B = \mathbf{a}_{B_1} = \mathbf{a}_{B_2} = \mathbf{a}_A + \boldsymbol{\alpha}_1 \times \mathbf{r}_B + \boldsymbol{\omega}_1 \times (\boldsymbol{\omega}_1 \times \mathbf{r}_B) = \boldsymbol{\alpha}_1 \times \mathbf{r}_B - \omega_1^2 \mathbf{r}_B$$
$$= -\omega_1^2 \mathbf{r}_B = -(-5.236)^2 (0.087 \boldsymbol{\imath} + 0.050 \mathbf{J}) = -2.374 \, \boldsymbol{\imath} - 1.371 \, \mathbf{J} \ \text{m/s}^2.$$

The angular acceleration of link 1 is $\boldsymbol{\alpha}_1 = \dot{\boldsymbol{\omega}}_1 = \mathbf{0}$.

The MATLAB commands for the velocity and acceleration of $B_1 = B_2$ are

```
n = -50.; %rpm
omega1_ = [0 0 pi*n/30];
alpha1_ = [0 0 0];
vA_ = [0 0 0];
aA_ = [0 0 0];
vB1_ = vA_ + cross(omega1_,rB_);
vB2_ = vB1_;
aB1_ = aA_ ...
       + cross(alpha1_,rB_) - dot(omega1_,omega1_)*rB_;
aB2_ = aB1_;
fprintf...
('vB1_ = vB2_ =[%6.3f,%6.3f,%g] (m/s)\n',vB1_)
fprintf...
('aB1_ = aB2_ =[%6.3f,%6.3f,%g] (m/s^2)\n',aB1_)
```

The velocity of the point B_3 on the link 3 is calculated in terms of the velocity of the point B_2 on the link 2

$$\mathbf{v}_{B_3} = \mathbf{v}_{B_2} + \mathbf{v}_{B_{32}}^{rel} = \mathbf{v}_{B_2} + \mathbf{v}_{B_{32}}, \tag{2.1}$$

where $\mathbf{v}_{B_{32}}^{rel} = \mathbf{v}_{B_{32}}$ is the relative velocity of B_3 with respect to a reference frame attached to link 2. This relative velocity is parallel to the sliding direction BC, $\mathbf{v}_{B_{32}} \| BC$, or

$$\mathbf{v}_{B_{32}} = v_{B_{32}} \cos \phi_2 \mathbf{i} + v_{B_{32}} \sin \phi_2 \mathbf{j}, \tag{2.2}$$

where $\phi_2 = 53.794°$ is known from position analysis. The points B_3 and C are on the link 3 and

$$\mathbf{v}_{B_3} = \mathbf{v}_C + \boldsymbol{\omega}_3 \times \mathbf{r}_{CB} = \boldsymbol{\omega}_3 \times (\mathbf{r}_B - \mathbf{r}_C), \tag{2.3}$$

where $\mathbf{v}_C \equiv \mathbf{0}$ and the angular velocity of link 3 is

$$\boldsymbol{\omega}_3 = \omega_3 \mathbf{k}.$$

Equations (2.1), (2.2), and (2.3) give

$$\begin{vmatrix} \mathbf{i} & \mathbf{j} & \mathbf{k} \\ 0 & 0 & \omega_3 \\ x_B - x_C & y_B - y_C & 0 \end{vmatrix} = \mathbf{v}_{B_2} + v_{B_{32}} \cos \phi_2 \mathbf{i} + v_{B_{32}} \sin \phi_2 \mathbf{j}. \tag{2.4}$$

Equation (2.4) represents a vectorial equations with two scalar components on x-axis and y-axis and with two unknowns ω_3 and $v_{B_{32}}$

$$-\omega_3(y_B - y_C) = v_{Bx} + v_{B_{32}} \cos \phi_2,$$
$$\omega_3(x_B - x_C) = v_{By} + v_{B_{32}} \sin \phi_2,$$

or

$$-\omega_3(0.050) = 0.262 + v_{B_{32}} \cos 53.794°,$$
$$\omega_3(0.087 - 0.05) = -0.453 + v_{B_{32}} \sin 53.794°.$$

It results

$$\omega_3 = \omega_2 = -7.732 \text{ rad/s} \text{ and } v_{B_{32}} = 0.211 \text{ m/s}.$$

The vectorial Eq. (2.4) is obtained in MATLAB with:

```
omega3z=sym('omega3z','real');
vB32=sym('vB32','real');
omega3u_ = [ 0 0 omega3z];
% omega3z unknown (to be calculated)
% vB32 unknown (to be calculated)
vC_ = [0 0 0]; % C is fixed
% vB3_ = vC_ + omega3_ x rCB_
% (B3 & C are points on link 3)
vB3_ = vC_ + cross(omega3u_,rB_-rC_);
% point B2 is on link 2 and point B3 is on link 3
% vB3_ = vB2_ + vB3B2_
% between the links 2 and 3 there is a translational
% joint B_T
% vB3B2_ is the relative velocity of B3 wrt 2
% vB3B2_ is parallel to the sliding direcion BC
% vB3B2_ is written as a vector
vB3B2u_ = vB32*[cos(phi2) sin(phi2) 0];
% vB3 = vB2 + vB3B2
eqvB_ = vB3_ - vB2_ - vB3B2u_;
% vectorial equation
% the component of the vectorial equation on x-axis
eqvBx = eqvB_(1);
% the component of the vectorial equation on y-axis
eqvBy = eqvB_(2);
% two equations eqvBx and eqvBy with two unknowns
```

The two scalar equations are displayed with:

```
digits(3)
% print the equations for calculating omega3 and vB32
fprintf...
```

```
('vB3_ = vC_ + omega3_ x rCB_ = vB2_ + vB3B2_ => \n')
qvBx=vpa(eqvBx);
fprintf('x-axis: %s = 0 \n',char(qvBx))
qvBy=vpa(eqvBy);
fprintf('y-axis: %s = 0 \n',char(qvBy))
fprintf('=>\n')
```

The solutions of the system are obtained with:

```
% solve for omega3z and vB32
solvB = solve(eqvBx,eqvBy);
omega3zs=eval(solvB.omega3z);
vB32s=eval(solvB.vB32);
omega3_ = [0 0 omega3zs];
omega2_ = omega3_;
vB3B2_ = vB32s*[cos(phi2) sin(phi2) 0];
```

The vectorial results are:

```
omega2_ = omega3_ = [0,0,-7.732](rad/s)
vB32_ = [ 0.125, 0.170,0] (m/s)
```

The acceleration of the point B_3 on the link 3 is calculated in terms of the acceleration of the point B_2 on the link 2

$$\mathbf{a}_{B_3} = \mathbf{a}_{B_2} + \mathbf{a}^{rel}_{B_3 B_2} + \mathbf{a}^{cor}_{B_3 B_2} = \mathbf{a}_{B_2} + \mathbf{a}_{B_{32}} + \mathbf{a}^{cor}_{B_{32}}, \tag{2.5}$$

where $\mathbf{a}^{rel}_{B_3 B_2} = \mathbf{a}_{B_{32}}$ is the relative acceleration of B_3 with respect to B_2 on link 2. This relative acceleration is parallel to the sliding direction BC, $\mathbf{a}_{B_{32}} \| BC$, or

$$\mathbf{a}_{B_{32}} = a_{B_{32}} \cos \phi_2 \mathbf{1} + a_{B_{32}} \sin \phi_2 \mathbf{J}. \tag{2.6}$$

The Coriolis acceleration of B_3 realative to B_2 is

$$\mathbf{a}^{cor}_{B_{32}} = 2\,\boldsymbol{\omega}_3 \times \mathbf{v}_{B_{32}} = 2\,\boldsymbol{\omega}_2 \times \mathbf{v}_{B_{32}} = 2 \begin{vmatrix} \mathbf{1} & \mathbf{J} & \mathbf{k} \\ 0 & 0 & \omega_3 \\ v_{B_{32}} \cos \phi_2 & v_{B_{32}} \sin \phi_2 & 0 \end{vmatrix} =$$

$$2(-\omega_3 v_{B_{32}} \sin \phi_2 \mathbf{1} + \omega_3 v_{B_{32}} \cos \phi_2 \mathbf{J}) =$$

$$2[-(-7.732)(0.211) \sin 53.794° \mathbf{1} + (-7.732)(0.211) \cos 53.794° \mathbf{J}] =$$

$$2.636\mathbf{1} - 1.930\mathbf{J} \ \text{m/s}^2. \tag{2.7}$$

The points B_3 and C are on the link 3 and

$$\mathbf{a}_{B_3} = \mathbf{a}_C + \boldsymbol{\alpha}_3 \times \mathbf{r}_{CB} - \omega_3^2 \mathbf{r}_{CB}, \tag{2.8}$$

where $\mathbf{a}_C \equiv \mathbf{0}$ and the angular acceleration of link 3 is

$$\boldsymbol{\alpha}_3 = \alpha_3 \mathbf{k}.$$

Equations (2.5), (2.6), (2.7), and (2.8) give

$$\begin{vmatrix} \mathbf{I} & \mathbf{J} & \mathbf{k} \\ 0 & 0 & \alpha_3 \\ x_B - x_C & y_B - y_C & 0 \end{vmatrix} - \omega_3^2 (\mathbf{r}_B - \mathbf{r}_C) =$$

$$\mathbf{a}_{B_2} + a_{B_{32}}(\cos\phi_2\mathbf{I} + \sin\phi_2\mathbf{J}) + 2\,\boldsymbol{\omega}_3 \times \mathbf{v}_{B_{32}}. \tag{2.9}$$

Equation (2.9) represents a vectorial equations with two scalar components on x-axis and y-axis and with two unknowns α_3 and $a_{B_{32}}$

$$-\alpha_3(y_B - y_C) - \omega_3^2(x_B - x_C) = a_{Bx} + a_{B_{32}}\cos\phi_2 - 2\omega_3 v_{B_{32}}\sin\phi_2,$$
$$\alpha_3(x_B - x_C) - \omega_3^2(y_B - y_C) = a_{By} + a_{B_{32}}\sin\phi_2 + 2\omega_3 v_{B_{32}}\cos\phi_2,$$

or

$$-\alpha_3(0.05) - (-7.732)^2(0.087 - 0.050) =$$
$$-2.374 + a_{B_{32}}\cos 53.794° - 2(-7.732)(0.211)\sin 53.794°,$$
$$\alpha_3(0.087 - 0.050) - (-7.732)^2(0.050) =$$
$$-1.371 + a_{B_{32}}\sin 53.794° + 2(-7.732)(0.211)\cos 53.794°.$$

It results
$$\alpha_3 = \alpha_2 = -34.865 \text{ rad/s}^2 \quad \text{and} \quad a_{B_{32}} = -1.196 \text{ m/s}^2.$$

The MATLAB commands for calculating α_3 and $a_{B_{32}}$ are:

```
% Coriolis acceleration
aB3B2cor_ = 2*cross(omega3_,vB3B2_);
alpha3z=sym('alpha3z','real');
aB32=sym('aB32','real'); % aB32 unknown
alpha3u_ = [0 0 alpha3z]; % alpha3z unknown
aC_ = [0 0 0]; % C is fixed
% aB3 acceleration of B3
aB3_=...
    aC_+cross(alpha3u_,rB_-rC_)...
    -dot(omega3_,omega3_)*(rB_-rC_);
% aB3B2_ relative velocity of B3 wrt 2
% aB3B2_ parallel to the sliding direcion BC
aB3B2u_ = aB32*[cos(phi2) sin(phi2) 0];
% aB3_ = aB2_ + aB3B2_ + aB3B2cor_
```

```
eqaB_ = aB3_ - aB2_ - aB3B2u_ - aB3B2cor_;
% vectorial equation
eqaBx = eqaB_(1); % equation component on x-axis
eqaBy = eqaB_(2); % equation component on y-axis
solaB = solve(eqaBx,eqaBy);
alpha3zs=eval(solaB.alpha3z);
aB32s=eval(solaB.aB32);
alpha3_ = [0 0 alpha3zs];
alpha2_ = alpha3_;
aB32_ = aB32s*[cos(phi2) sin(phi2) 0];
```

and the vectorial solutions are:

```
alpha2_=alpha3_=[0,0,-34.865](rad/s^2)
aB32_=[-0.706,-0.965,-0] (m/s^2)
```

The velocity of D is

$$\mathbf{v}_D = \mathbf{v}_C + \boldsymbol{\omega}_3 \times \mathbf{r}_{CD} = \boldsymbol{\omega}_3 \times (\mathbf{r}_D - \mathbf{r}_C) =$$

$$\begin{vmatrix} \mathbf{\imath} & \mathbf{\jmath} & \mathbf{k} \\ 0 & 0 & \omega_3 \\ x_D - x_C & y_D - y_C & 0 \end{vmatrix} = \begin{vmatrix} \mathbf{\imath} & \mathbf{\jmath} & \mathbf{k} \\ 0 & 0 & -7.732 \\ 0.139 - 0.050 & 0.121 & 0 \end{vmatrix} =$$

$$0.936\,\mathbf{\imath} - 0.685\,\mathbf{\jmath} \text{ m/s}.$$

The acceleration of D is

$$\mathbf{a}_D = \mathbf{a}_C + \boldsymbol{\alpha}_3 \times \mathbf{r}_{CD} - \omega_3^2 \mathbf{r}_{CD} = \boldsymbol{\alpha}_3 \times (\mathbf{r}_D - \mathbf{r}_C) - \omega_3^2(\mathbf{r}_D - \mathbf{r}_C) =$$

$$\begin{vmatrix} \mathbf{\imath} & \mathbf{\jmath} & \mathbf{k} \\ 0 & 0 & \alpha_3 \\ x_D - x_C & y_D - y_C & 0 \end{vmatrix} - \omega_3^2 [(x_D - x_C)\mathbf{\imath} + (y_D - y_C)\mathbf{\jmath}] =$$

$$\begin{vmatrix} \mathbf{\imath} & \mathbf{\jmath} & \mathbf{k} \\ 0 & 0 & -34.865 \\ 0.139 - 0.050 & 0.121 & 0 \end{vmatrix} -$$

$$(-7.732)^2 [(0.139 - 0.050)\,\mathbf{\imath} + (0.121)\,\mathbf{\jmath}] =$$

$$- 1.077\,\mathbf{\imath} - 10.324\,\mathbf{\jmath} \text{ m/s}^2.$$

The velocity and acceleration of D are calculated in MATLAB with:

```
vD_ = vC_ + cross(omega3_,rD_-rC_);
aD_=...
    aC_+cross(alpha3_,rD_-rC_)...
    -dot(omega3_,omega3_)*(rD_-rC_);
```

The position and acceleration vectors of the centers of mass C1, C2, C3 of each link are calculated with:

```
rC1_ = rB_/2;
rC2_ = rB_;
rC3_ = (rD_+rC_)/2;
aC1_ = aB1_/2;
aC2_ = aB1_;
aC3_ = (aD_+aC_)/2;
```

and the results are:

```
rC1_ = [ 0.043,  0.025,0]  (m)
rC2_ = [ 0.087,  0.050,0]  (m)
rC3_ = [ 0.094,  0.061,0]  (m)
aC1_ = [-1.187,-0.685,0]   (m/s^2)
aC2_ = [-2.374,-1.371,0]   (m/s^2)
aC3_ = [-0.538,-5.162,0]   (m/s^2)
```

2.3   Dynamic Force Analysis

The input data for the geometry of the links are:

```
h = 0.01; % height of the links
d = 0.001; % depth of the links
hSlider = 0.02; % height of the slider
wSlider = 0.04; % width of the slider
rho = 8000; % density of the material
g = 9.807; % gravitational acceleration
```

The mass of the link 1, m_1, and the mass moment of inertia, I_{C_1}, are calculated with

$$m_1 = \rho\, AB\, h\, d,$$
$$I_{C_1} = m_1(AB^2 + h^2)/12.$$

The force of gravity, $\mathbf{G}_1$, the force of inertia, $\mathbf{F}_{in1}$, and the moment of inertia, $\mathbf{M}_{in1}$, of the link 1 are calculated using the MATAB commands:

```
m1 = rho*AB*h*d; % mass
G1_ = [0,-m1*g,0]; % force of gravity
Fin1_ = -m1*aC1_; % force of inertia
IC1 = m1*(AB^2+h^2)/12; % mass moment of inertia
IA  = IC1+m1*(AB/2)^2; % A fixed point
```

```
alpha1_ = [0 0 0]; % angular acceleration
Min1_ = -IC1*alpha1_; % moment of inertia
```

For the links 2 and 3 the MATLAB statements for the force of gravity, the force of inertia, and the moment of inertia are:

```
m2 = rho*hSlider*wSlider*d;
G2_ = [0,-m2*g,0];
Fin2_ = -m2*aC2_;
IC2 = m2*(hSlider^2+wSlider^2)/12;
Min2_ = -IC2*alpha2_;

m3 = rho*CD*h*d;
G3_ = [0,-m3*g,0];
Fin3_ = -m3*aC3_;
IC3 = m3*(CD^2+h^2)/12;
IC  = IC3+m3*(CD/2)^2; % C fixed point
Min3_ = -IC3*alpha3_;
```

The results are:

```
Link 1
m1 =  0.008 (kg)
G1_ = - m1 g_ = [0,-0.078,0] (N)
m1 aC1_ = [-0.009,-0.005,0] (N)
Fin1_= - m1 aC1_ = [ 0.009, 0.005,0] (N)
IC1 = 6.73333e-06 (kg m^2)
IC1 alpha1_=[ 0.000, 0.000,0](N m)
Min1_=-IC1 alpha1_=[0, 0, 0](N m)

Link 2
m2 =  0.006 (kg)
G2_ = - m2 g_ = [0,-0.063,0] (N)
m2 aC2_ = [-0.015,-0.009,0] (N)
Fin2_= - m2 aC2_ =[ 0.015, 0.009,0] (N)
IC2 = 1.067e-06 (kg m^2)
IC2 alpha2_=[0,0,-3.719e-05](N m)
Min2_=-IC2 alpha2_=[0,0,3.719e-05](N m)

Link 3
m3 =  0.012 (kg)
G3_ = - m3 g_ = [0, -0.118, 0] (N)
m3 aC3_ = [-0.006,-0.062,0] (N)
Fin3_ = - m3 aC3_ =[ 0.006, 0.062,0] (N)
IC3 = 2.260e-05 (kg m^2)
```

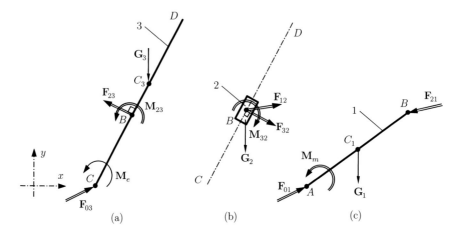

Fig. 2.5 Free body diagrams for the links of the R-RTR mechanism

```
IC3 alpha3_=[0,0,-7.880e-04](N m)
Min3_=-IC3 alpha3_=[0,0,7.880e-04](N m)
```

The external moment on link 3 is calculated with:

```
Me=100; % N m
Me_ = -sign(omega3_(3))*[0,0,Me];
```

The joint forces are calculated using Newton-Euler equations of motion for each link. The free body diagrams for the links of the R-RTR mechanism are shown in Fig. 2.5.

The dynamic force analysis starts with the last link 3 because the given external moment acts on this link. For link 3 the unknowns are:

```
F23x=sym('F23x','real'); % force of 2 on 3 at B x-comp
F23y=sym('F23y','real'); % force of 2 on 3 at B y-comp
M23z=sym('M23z','real'); % moment of 2 on 3 z-component
F23_=[F23x, F23y, 0]; % unknown joint force
M23_=[0, 0, M23z]; % unknown joint moment
```

Because the joint between 2 and 3 at B is a translational joint the reaction force F23_ is perpendicular to the sliding direction BC:

```
% F23_ perpendicular to BC: F32.BC = 0 =>
eqF23 = (rB_-rC_)*F23_.'; % (1)
```

A moment equation for all the forces and moments that act on link 3 with respect to the fixed point C can be written:

```
% sum M of 3 wrt C (C 'fixed' point)
% CB_ x F23_ + CC3_ x G3_ + M23_ + Me_ = IC*alpha3_
eqM3C_ = ...
  cross(rB_-rC_,F23_)+cross(rC3_-rC_,G3_)...
  +M23_+Me_-IC*alpha3_;
eqM3z = eqM3C_(3); % (2)
```

There are two scalar equations %(1) and %(2) with three unknowns F23x, F23y, M23z.

The force calculation will continue with link 2. For link 2 a new unknown force, the reaction of link 1 on link 2 at *B*, is introduced:

```
F12x=sym('F12x','real'); % force of 1 on 2 at B x-comp
F12y=sym('F12y','real'); % force of 1 on 2 at B y-comp
F12_=[F12x, F12y, 0]; % unknown joint force
```

The Newton-Euler equations of motion for link 2 are written as:

```
% sum F of 2
% F12_+G2_+(-F23_) = m2*aC2_
eqF2_ = F12_+G2_+(-F23_)-m2*aC2_;
eqF2x = eqF2_(1); % (3)
eqF2y = eqF2_(2); % (4)

% sum M of 2 wrt B=C2
% -M23_=IC2*alpha2_
eqM2_ = -M23_ - IC2*alpha2_;
eqM2z = eqM2_(3); % (5)
```

Now there are 5 equations with 5 unknowns and the system can be solved using MATLAB:

```
% Eqs.(1)-(5) =>
sol23=solve(eqF23,eqM3z,eqF2x,eqF2y,eqM2z);
F23xs=eval(sol23.F23x);
F23ys=eval(sol23.F23y);
M23zs=eval(sol23.M23z);
F12xs=eval(sol23.F12x);
F12ys=eval(sol23.F12y);

F12s_=[F12xs, F12ys, 0];
F23s_=[F23xs, F23ys, 0];
M23s_=[0,0,M23zs];
```

The results are:

```
F23_ = [1302.143,-953.235,0] (N)
F12_ = [1302.128,-953.181,0] (N)
M23_ = [0,0,0.000037] (N m)
```

The reaction force of the ground 0 on link 3 at C is:

```
% F03_+G3_+F23_ = m3*aC3_ =>
F03_ = m3*aC3_-(G3_+F23s_);
```

and the numerical value is:

```
F03_ = [-1302.149,953.291,0] (N)
```

The motor moment (equilibrium moment), Mm_, required for the dynamic equilibrium of the mechanism is determined from:

```
% sum M of 1 wrt A (A 'fixed' point)
% IA*alpha1_ = rC1_ x G1_ + rB_ x (-F12s_) + Mm_ =>
Mm_=IA*alpha1_-(cross(rC1_,G1_)+cross(rB_,-F12s_));
```

and the result is:

```
Mm_ = [0,0,-147.651] (N m)
```

Derivative Method for Velocity and Acceleration

The velocities of the mechanical system are calculated taking the derivatives of the positions and the accelerations are obtained with the derivatives of the velocities. For the planar R-RTR mechanism in Fig. 2.1 the following data are given:

```
AB = 0.10 ;   % (m)
AC = 0.05 ;   % (m)
CD = 0.15 ;   % (m)
% position angle of driver 1
phi1 = pi/6; % (rad)
% angular speed of driver 1
n = -50;      % (rpm)
omega1 = pi*n/30; % (rad/s) angular velocity
alpha1 = 0;       % (rad/s^2) angular speed

% A(xA, yA, 0) origin
xA = 0; yA = 0;
rA_ = [xA yA 0];
```

```
% C(xC, yC, 0) position of joint at C
xC = AC ; yC = 0;
rC_ = [xC yC 0];
```

For the position angle of the driver link the symbolic variable phi(t) is introduced:

```
syms phi(t)
% phi(t) the angle of the diver link with the horizontal
% phi(t) is a function of time, t
```

The coordinates of the joint B are:

```
xB = AB*cos(phi(t));
yB = AB*sin(phi(t));
% position vector of B in terms of phi(t) - symbolic
rB_ = [xB yB 0];
```

The numerical values for the joint B are:

```
xBn = subs(xB, phi(t), phi1); % xB for phi(t)=phi1=pi/6
yBn = subs(yB, phi(t), phi1); % yB for phi(t)=phi1=pi/6
rBn_ = subs(rB_, phi(t), phi1); % rB_ for phi(t)=phi1=pi/6
% subs(expr,lhs,rhs) replaces lhs with rhs in symbolic expression
fprintf('rB_ =  [ %6.3f, %6.3f, %g ] (m)\n', rBn_)
% rB_ =  [  0.087,  0.050, 0 ] (m)
```

The linear velocity vector of $B_1 = B_2$ is

$$\mathbf{v}_B = \mathbf{v}_{B_1} = \mathbf{v}_{B_2} = \frac{dx_B}{dt}\mathbf{1} + \frac{dy_B}{dt}\mathbf{J} = \dot{x}_B\mathbf{1} + \dot{y}_B\mathbf{J},$$

or in MATLAB

```
% vB_=vB1_=vB2_ in terms of phi(t) and diff(phi(t),t)
vB_ = diff(rB_, t); % velocity of B1=B2
% diff(S,t) differentiates S with respect to t
```

For the numerical values first replace diff(phi(t), t, 2) with alpha1, then replace diff(phi(t), t) with omega1, and then replace phi(t) with phi1 . A list with the symbolical variables phi(t), diff(phi(t), t), and diff(phi(t),t,2) and another list with numerical values phi1, omega1, and alpha1 are introduced:

```
% list for symbolical variables  phi''(t), phi'(t), phi(t)
slist = {diff(phi(t),t,2), diff(phi(t),t), phi(t)};
% list for numerical values
nlist = {alpha1, omega1, phi1}; % numerical values for slist
```

```
% diff(phi(t), t, 2) -> aplha1 = 0
% diff(phi(t), t) -> omega1
% phi(t) -> phi1 = pi/6
```

The numerical value for the symbolic vector vB_ are:

```
vBn_ = subs(vB_, slist, nlist); % replaces slist with nlist in vB_
fprintf('vB1_ = vB2_ = [ %6.3f, %6.3f, %g ] (m/s)\n', vBn_)
% norm(v_) is the magnitude of a vector v_
vBn = norm(vBn_);
fprintf('|vB1| = |vB2| = %6.3f (m/s) \n', vBn)
%vB1_ = vB2_ = [  0.262, -0.453, 0 ] (m/s)
%|vB1| = |vB2| =  0.524 (m/s)
```

The linear acceleration vector of $B_1 = B_2$ is:

```
aB_  = diff(vB_, t);
% numerical value for aB
aBn_ = subs(aB_,slist,nlist);
fprintf('aB1_ = aB2_ = [ %6.3f, %6.3f, %g ] (m/s^2)\n', aBn_)
aBn = norm(aBn_);
fprintf('|aB1| = |aB2| = %6.3f (m/s^2) \n', aBn)
%aB1_ = aB2_ = [ -2.374, -1.371, 0 ] (m/s^2)
%|aB1| = |aB2| =  2.742 (m/s^2)
```

The angle of the link 2 (link 3) with the horizontal, $\phi_2(t) = \phi_3(t)$, is determined as a function of time t from the equation of the slope of the line BC:

```
phi2 = atan((yB-yC)/(xB-xC));   % phi2 in terms of phi(t)
```

where atan(z) gives the arc tangent of the number z. The numerical value is:

```
phi2n = subs(phi2, phi(t), phi1); % phi2 for phi(t)=pi/6
fprintf('phi2 = phi3 = %6.3f (degrees) \n', phi2n*180/pi)
% phi2 = phi3 = 53.794 (degrees)
```

The angular velocity and angular acceleration of link 2 and link 3 are:

```
% omega2 in terms of phi(t) and diff('phi(t)',t)
dphi2 = diff(phi2, t);
dphi2nn = subs(dphi2, diff(phi(t), t), omega1);
% numerical value for omega2
dphi2n  = subs(dphi2nn, phi(t), phi1);
fprintf('omega2 = omega3 = %6.3f (rad/s) \n', dphi2n)

% alpha2 in terms of phi(t), diff('phi(t)',t), and diff('phi(t)',t,2)
ddphi2  = diff(dphi2, t);
% numerical value for alpha2
ddphi2n = subs(ddphi2, slist, nlist);
```

```
fprintf('alpha2 = alpha3 = %6.3f (rad/s^2) \n', ddphi2n)

%omega2 = omega3 = -7.732 (rad/s)
%alpha2 = alpha3 = -34.865 (rad/s^2)
```

The position of the point D is calculated from the following equations

$$[x_D(t) - x_C]^2 + [y_D(t) - y_C]^2 = CD^2,$$
$$\frac{y_B(t) - y_C}{x_B(t) - x_C} = \frac{y_D(t) - y_C}{x_D(t) - x_C},$$

or in MATLAB:

```
syms xDsol yDsol
eqnD1 = (xDsol - xC)^2 + (yDsol - yC)^2-CD^2 ;
eqnD2 = (yB-yC)/(xB-xC)-(yDsol-yC)/(xDsol-xC);
solD = solve(eqnD1, eqnD2, xDsol, yDsol);
xDpositions = solD.xDsol;
yDpositions = solD.yDsol;
xD1 = xDpositions(1); xD2 = xDpositions(2);
yD1 = yDpositions(1); yD2 = yDpositions(2);
```

Two sets of solutions are found for point D that are functions of the angle $\phi(t)$. The correct position of point D for the given angle of the driver link 1 can be selected with the condition:

```
% select the correct position for D for the given input angle
xD1n = subs(xD1, phi(t), phi1); % xD1 for phi(t)=pi/6
if xD1n > xC
    xD = xD1; yD = yD1;
else
    xD = xD2; yD = yD2;
end
% position vector of D in term of phi(t) - symbolic
rD_ = [ xD yD 0 ];
xDn = subs(xD, phi(t), phi1); % xD for phi(t)=pi/6
yDn = subs(yD, phi(t), phi1); % yD for phi(t)=pi/6
rDn_ = [ xDn yDn 0 ]; % rD_ for phi(t)=pi/6
fprintf('rD_ =   [ %6.3f, %6.3f, %g ] (m)\n', rDn_)
% rD_ =   [  0.139,  0.121, 0 ] (m)
```

The linear velocity and acceleration vectors of D are:

```
% velocity of D
% vD_ in terms of phi(t) and diff(phi(t),t)
vD_  = diff(rD_,t);
```

```
% numerical value for vD_
vDn = subs(vD_, slist, nlist);
fprintf('vD_  = [ %6.3f, %6.3f, %g ] (m/s)\n', vDn)
fprintf('|vD| = %6.3f (m/s) \n', norm(vDn))

% acceleration of D_
% aD in terms of phi(t), diff(phi(t),t), and diff(phi(t),t,2)
aD_  = diff(vD_, t);
% numerical value for aD_
aDn_ = subs(aD_, slist, nlist);
fprintf('aD_  = [ %6.3f, %6.3f, %g ] (m/s^2)\n', aDn_)
fprintf('|aD| = %6.3f (m/s^2) \n', norm(aDn_))

%vD_  = [   0.936, -0.685,  0 ] (m/s)
%|vD| =   1.160 (m/s)
%aD_  = [ -1.077, -10.324,  0 ] (m/s^2)
%|aD| = 10.380 (m/s^2)
```

2.4 Problems

2.1 The R-RRT mechanism has homogeneous slender links AB and BC as shown in
Fig. 2.6. The masses of the links are m_1, m_2, m_3. The gravitational acceleration
is g. The driver link 1 rotates with a constant speed of n. Find all the joint
reaction forces and the motor (equilibrium) moment on the driver link at the
instant shown. For $\phi_1 = \phi = 0°$ the following data are given

Fig. 2.6 R-RRT mechanism

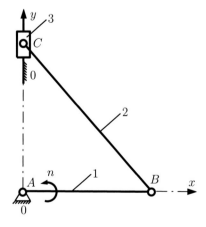

```
AB = 1;  % (m)
BC = 2;  % (m)
phi = 0;% (rad) driver link 1 angle
n = 240/pi;  % (rpm) driver link 1 angular speed
% external force Fe_ on link 3 (N)
fe = 100;  % (N)
% Fe_ = -sign(vC)*[0 fe 0];
m1 = 1;  % (kg) mass of link 1
% mass moment of inertia of link 1 wrt center of mass C1
IC1 = m1*(AB^2)/12;
m2 = 2;  % (kg) mass of link 2
% mass moment of inertia of link 2 wrt center of mass C2
IC2 = m2*(BC^2)/12;
m3 = 1;  % (kg) mass of slider 1
hSlider = 0.01;  % (m) height of the slider
wSlider = 0.01;  % (m) depth of the slider
% mass moment of inertia of link 3 wrt center of mass C3
IC3 = m3*(hSlider^2+wSlider^2)/12;
g = 9.81;  % gravitational acceleration (m/s^2)
```

2.2 The R-RRR mechanism shown in Fig. 2.7 has the dimensions: $AB = BC = CD = 1$ m. The driver link 1 rotates with a constant speed of $\omega = \omega_1 = 1$ rad/s. The dynamic of the mechanism is considered at the instant shown when the driver link 1 makes an angle $\phi = \phi_1 = \pi/4$ rad with the horizontal axis. The mass of the slender links 1, 2, and 3 are $m_1 = m_2 = m_3 = 1$ kg. The height of the links $h = 0.015$ m. The depth of the links is $d = 0.01$ m. The density of the material $\rho = 8000$ kg/m³. The external moment on link 3 is $\mathbf{M}_{3ext} = -10 \mathbf{k}$ N m. The gravitational acceleration is $g = 9.81$ m/s². Determine the moment $\mathbf{M}$ that acts on link 1 required for dynamic equilibrium and the joint reaction forces.

2.3 The R-RTR mechanism shown in Fig. 2.8 has the following dimensions: $AB = 1$ m, $AC = \sqrt{2}$ m, $BD = 3$ m. The driver link 1 is rotating with a constant angular speed $\omega = 2$ rad/s. The mechanism is analyzed for $\phi = \phi_1 = 45°$. The height of links 1 and 2 is $h = 0.01$ m. The width of the slider 3 is $w_{Slider} = 0.01$ m and the height is $h_{Slider} = 0.01$ m. The depth of the links is $d = 0.01$ m. The density of the material $\rho = 8000$ kg/m³. There is an external moment on link 3

$$\mathbf{M}_{3ext} = -100 \, \mathrm{sign}(\omega_3) \, \mathbf{k} \quad \text{N m}.$$

Fig. 2.7 R-RRR mechanism

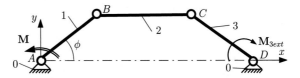

Fig. 2.8 R-RTR mechanism

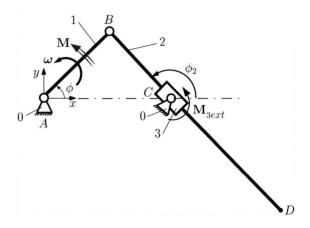

Determine the moment **M** that acts on link 1 required for dynamic equilibrium and the joint forces for the mechanism for $\phi = \phi_1 = 45°$.

References

1. E.A. Avallone, T. Baumeister, A. Sadegh, *Marks' Standard Handbook for Mechanical Engineers*, 11th edn. (McGraw-Hill Education, New York, 2007)
2. I.I. Artobolevski, *Mechanisms in Modern Engineering Design* (MIR, Moscow, 1977)
3. M. Atanasiu, *Mechanics (Mecanica)* (EDP, Bucharest, 1973)
4. H. Baruh, *Analytical Dynamics* (WCB/McGraw-Hill, Boston, 1999)
5. A. Bedford, W. Fowler, *Dynamics* (Addison Wesley, Menlo Park, CA, 1999)
6. A. Bedford, W. Fowler, *Statics* (Addison Wesley, Menlo Park, CA, 1999)
7. F.P. Beer et al., *Vector Mechanics for Engineers: Statics and Dynamics* (McGraw-Hill, New York, NY, 2016)
8. M. Buculei, D. Bagnaru, G. Nanu, D.B. Marghitu, *Computing Methods in the Analysis of the Mechanisms with Bars* (Scrisul Romanesc, Craiova, 1986)
9. M.I. Buculei, *Mechanisms* (University of Craiova Press, Craiova, Romania, 1976)
10. D. Bolcu, S. Rizescu, *Mecanica* (EDP, Bucharest, 2001)
11. J. Billingsley, *Essential of Dynamics and Vibration* (Springer, 2018)
12. R. Budynas, K.J. Nisbett, *Shigley's Mechanical Engineering Design*, 9th edn. (McGraw-Hill, New York, 2013)
13. J.A. Collins, H.R. Busby, G.H. Staab, *Mechanical Design of Machine Elements and Machines* (2nd edn., Wiley, 2009)
14. M. Crespo da Silva, *Intermediate Dynamics for Engineers* (McGraw-Hill, New York, 2004)
15. A.G. Erdman, G.N. Sandor, *Mechanisms Design* (Prentice-Hall, Upper Saddle River, NJ, 1984)
16. M. Dupac, D.B. Marghitu, *Engineering Applications: Analytical and Numerical Calculation with MATLAB* (Wiley, Hoboken, NJ, 2021)
17. A. Ertas, J.C. Jones, *The Engineering Design Process* (Wiley, New York, 1996)
18. F. Freudenstein, An application of Boolean Algebra to the Motion of Epicyclic Drivers. J. Eng. Ind. 176–182 (1971)
19. J.H. Ginsberg, *Advanced Engineering Dynamics* (Cambridge University Press, Cambridge, 1995)

20. H. Goldstein, *Classical Mechanics* (Addison-Wesley, Redwood City, CA, 1989)
21. D.T. Greenwood, *Principles of Dynamics* (Prentice-Hall, Englewood Cliffs, NJ, 1998)
22. A.S. Hall, A.R. Holowenko, H.G. Laughlin, *Schaum's Outline of Machine Design* (McGraw-Hill, New York, 2013)
23. B.G. Hamrock, B. Jacobson, S.R. Schmid, *Fundamentals of Machine Elements* (McGraw-Hill, New York, 1999)
24. R.C. Hibbeler, *Engineering Mechanics: Dynamics* (Prentice Hall, 2010)
25. T.E. Honein, O.M. O'Reilly, On the Gibbs–Appell equations for the dynamics of rigid bodies. J. Appl. Mech. **88**/074501-1 (2021)
26. R.C. Juvinall, K.M. Marshek, *Fundamentals of Machine Component Design*, 5th edn. (Wiley, New York, 2010)
27. T.R. Kane, *Analytical Elements of Mechanics*, vol. 1 (Academic Press, New York, 1959)
28. T.R. Kane, *Analytical Elements of Mechanics*, vol. 2 (Academic Press, New York, 1961)
29. T.R. Kane, D.A. Levinson, The use of Kane's dynamical equations in robotics. MIT Int. J. Robot. Res. **3**, 3–21 (1983)
30. T.R. Kane, P.W. Likins, D.A. Levinson, *Spacecraft Dynamics* (McGraw-Hill, New York, 1983)
31. T.R. Kane, D.A. Levinson, *Dynamics* (McGraw-Hill, New York, 1985)
32. K. Lingaiah, *Machine Design Databook*, 2nd edn. (McGraw-Hill Education, New York, 2003)
33. N.I. Manolescu, F. Kovacs, A. Oranescu, *The Theory of Mechanisms and Machines* (EDP, Bucharest, 1972)
34. D.B. Marghitu, *Mechanical Engineer's Handbook* (Academic Press, San Diego, CA, 2001)
35. D.B. Marghitu, M.J. Crocker, *Analytical Elements of Mechanisms* (Cambridge University Press, Cambridge, 2001)
36. D.B. Marghitu, *Kinematic Chains and Machine Component Design* (Elsevier, Amsterdam, 2005)
37. D.B. Marghitu, *Mechanisms and Robots Analysis with MATLAB* (Springer, New York, NY, 2009)
38. D.B. Marghitu, M. Dupac, *Advanced Dynamics: Analytical and Numerical Calculations with MATLAB* (Springer, New York, NY, 2012)
39. D.B. Marghitu, M. Dupac, H.M. Nels, *Statics with MATLAB* (Springer, New York, NY, 2013)
40. D.B. Marghitu, D. Cojocaru, *Advances in Robot Design and Intelligent Control* (Springer International Publishing, Cham, Switzerland, 2016), pp. 317–325
41. D.J. McGill, W.W. King, *Engineering Mechanics: Statics and an Introduction to Dynamics* (PWS Publishing Company, Boston, 1995)
42. J.L. Meriam, L.G. Kraige, *Engineering Mechanics: Dynamics* (Wiley, New York, 2007)
43. C.R. Mischke, Prediction of stochastic endurance strength. Trans. ASME J. Vib. Acoust. Stress Reliab. Des. **109**(1), 113–122 (1987)
44. L. Meirovitch, *Methods of Analytical Dynamics* (Dover, 2003)
45. R.L. Mott, *Machine Elements in Mechanical Design* (Prentice Hall, Upper Saddle River, NJ, 1999)
46. W.A. Nash, *Strength of Materials* (Schaum's Outline Series, McGraw-Hill, New York, 1972)
47. R.L. Norton, *Machine Design* (Prentice-Hall, Upper Saddle River, NJ, 1996)
48. R.L. Norton, *Design of Machinery* (McGraw-Hill, New York, 1999)
49. O.M. O'Reilly, *Intermediate Dynamics for Engineers Newton-Euler and Lagrangian Mechanics* (Cambridge University Press, UK, 2020)
50. O.M. O'Reilly, *Engineering Dynamics: A Primer* (Springer, NY, 2010)
51. W.C. Orthwein, *Machine Component Design* (West Publishing Company, St. Paul, 1990)
52. L.A. Pars, *A Treatise on Analytical Dynamics* (Wiley, New York, 1965)
53. F. Reuleaux, *The Kinematics of Machinery* (Dover, New York, 1963)
54. D. Planchard, M. Planchard, *SolidWorks 2013 Tutorial with Video Instruction* (SDC Publications, 2013)
55. I. Popescu, *Mechanisms* (University of Craiova Press, Craiova, Romania, 1990)
56. I. Popescu, C. Ungureanu, *Structural Synthesis and Kinematics of Mechanisms with Bars* (Universitaria Press, Craiova, Romania, 2000)

57. I. Popescu, L. Luca, M. Cherciu, D.B. Marghitu, *Mechanisms for Generating Mathematical Curves* (Springer Nature, Switzerland, 2020)
58. C.A. Rubin, *The Student Edition of Working Model* (Addison-Wesley Publishing Company, Reading, MA, 1995)
59. J. Ragan, D.B. Marghitu, Impact of a Kinematic link with MATLAB and SolidWorks. Applied Mechanics and Materials **430**, 170–177 (2013)
60. J. Ragan, D.B. Marghitu, MATLAB dynamics of a free link with elastic impact, in *International Conference on Mechanical Engineering, ICOME 2013*, 16–17 May 2013 (Craiova, Romania, 2013)
61. J.C. Samin, P. Fisette, *Symbolic Modeling of Multibody Systems* (Kluwer, 2003)
62. A.A. Shabana, *Computational Dynamics* (Wiley, New York, 2010)
63. I.H. Shames, *Engineering Mechanics—Statics and Dynamics* (Prentice-Hall, Upper Saddle River, NJ, 1997)
64. J.E. Shigley, C.R. Mischke, *Mechanical Engineering Design* (McGraw-Hill, New York, 1989)
65. J.E. Shigley, C.R. Mischke, R.G. Budynas, *Mechanical Engineering Design*, 7th edn. (McGraw-Hill, New York, 2004)
66. J.E. Shigley, J.J. Uicker, *Theory of Machines and Mechanisms* (McGraw-Hill, New York, 1995)
67. D. Smith, *Engineering Computation with MATLAB* (Pearson Education, Upper Saddle River, NJ, 2008)
68. R.W. Soutas-Little, D.J. Inman, *Engineering Mechanics: Statics and Dynamics* (Prentice-Hall, Upper Saddle River, NJ, 1999)
69. J. Sticklen, M.T. Eskil, *An Introduction to Technical Problem Solving with MATLAB* (Great Lakes Press, Wildwood, MO, 2006)
70. A.C. Ugural, *Mechanical Design* (McGraw-Hill, New York, 2004)
71. R. Voinea, D. Voiculescu, V. Ceausu, *Mechanics (Mecanica)* (EDP, Bucharest, 1983)
72. K.J. Waldron, G.L. Kinzel, *Kinematics, Dynamics, and Design of Machinery* (Wiley, New York, 1999)
73. J.H. Williams Jr., *Fundamentals of Applied Dynamics* (Wiley, New York, 1996)
74. C.E. Wilson, J.P. Sadler, *Kinematics and Dynamics of Machinery* (Harper Collins College Publishers, New York, 1991)
75. H.B. Wilson, L.H. Turcotte, D. Halpern, *Advanced Mathematics and Mechanics Applications Using MATLAB* (Chapman & Hall/CRC, 2003)
76. S. Wolfram, *Mathematica* (Wolfram Media/Cambridge University Press, Cambridge, 1999)
77. National Council of Examiners for Engineering and Surveying (NCEES), *Fundamentals of Engineering. Supplied-Reference Handbook* (Clemson, SC, 2001)
78. * * *, *The Theory of Mechanisms and Machines* (Teoria mehanizmov i masin) (Vassaia scola, Minsk, Russia, 1970)
79. eCourses—University of Oklahoma. http://ecourses.ou.edu/home.htm
80. https://www.mathworks.com
81. http://www.eng.auburn.edu/~marghitu/
82. https://www.solidworks.com
83. https://www.wolfram.com

Chapter 3
Contour Analysis of a Mechanism with One Dyad

Abstract Closed contour equations for velocities and accelerations are given for closed kinematic mechanisms. The contour equations are based on contour diagrams and take into consideration the relative velocities and accelerations between two adjacent links. The inverse dynamics is based on contour diagrams and D'Alembert principle.

3.1 Closed Contour Equations

This study aims at providing an algebraic method to compute the velocities of a closed kinematic chain [3, 18, 71]. The method of independent contour (loop) equations is very efficient and can be applied to planar and spatial mechanical systems.

Consider two rigid bodies (j) and (k) connected by a joint or kinematic pair at A. The following relation exists between the velocity $\mathbf{v}_{A_j}$ of the point A_j and the velocity $\mathbf{v}_{A_k}$ of the point A_k

$$\mathbf{v}_{A_j} = \mathbf{v}_{A_k} + \mathbf{v}^r_{A_{jk}}, \tag{3.1}$$

where $\mathbf{v}^r_{A_{jk}} = \mathbf{v}^r_{A_j A_k} = \mathbf{v}_{A_j A_k}$ is the velocity of A_j as seen by an observer at A_k attached to body (k) or the relative velocity of A_j with respect to A_k, allowed at the joint A. The accelerations of A_j and A_k are expressed as

$$\mathbf{a}_{A_j} = \mathbf{a}_{A_k} + \mathbf{a}^r_{A_{jk}} + \mathbf{a}^c_{A_{jk}}, \tag{3.2}$$

where $\mathbf{a}^r_{A_{jk}} = \mathbf{a}^r_{A_j A_k} = \mathbf{a}_{A_j A_k}$ is the relative acceleration of A_j with respect to A_k and $\mathbf{a}^c_{A_{jk}} = \mathbf{a}^c_{A_j A_k}$ is the Coriolis acceleration given by

$$\mathbf{a}^c_{A_{jk}} = 2\,\boldsymbol{\omega}_k \times \mathbf{v}^r_{A_{jk}}, \tag{3.3}$$

where $\boldsymbol{\omega}_k$ is the angular velocity of the body (k). Equations (3.1) and (3.2) are useful even for coincident points belonging to two rigid bodies that may not be directly

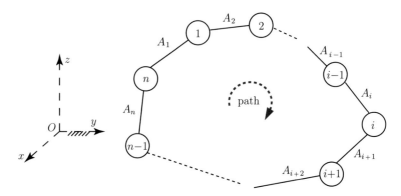

Fig. 3.1 Monoloop of a closed kinematic chain

connected. Figure 3.1 shows a monoloop closed kinematic chain with n rigid links. The joint A_i where $i = 1, 2, ..., n$ is the connection between the links (i) and $(i - 1)$. At the joint A_i there are two instantaneously coincident points. The point $A_{i,i}$ belongs to link (i), $A_{i,i} \in (i)$ and the point $A_{i,i-1}$ belongs to body $(i - 1)$, $A_{i,i-1} \in (i - 1)$. The absolute angular velocity of the rigid body (i) is

$$\boldsymbol{\omega}_i = \boldsymbol{\omega}_{i-1} + \boldsymbol{\omega}_{i,i-1}, \tag{3.4}$$

where $\boldsymbol{\omega}_{i,i-1}$ is the relative angular velocity of the rigid body (i) with respect to the rigid body $(i - 1)$. For the n link closed kinematic chain the equations for the angular velocities are

$$\boldsymbol{\omega}_1 = \boldsymbol{\omega}_n + \boldsymbol{\omega}_{1,n}; \ \boldsymbol{\omega}_2 = \boldsymbol{\omega}_1 + \boldsymbol{\omega}_{2,1}; \ ... \ \boldsymbol{\omega}_i = \boldsymbol{\omega}_{i-1} + \boldsymbol{\omega}_{i,i-1}; \ ... \ \boldsymbol{\omega}_n = \boldsymbol{\omega}_{n-1} + \boldsymbol{\omega}_{n,n-1}. \tag{3.5}$$

Summing the expressions in Eq. (3.5) the following relation is obtained

$$\boldsymbol{\omega}_{1,n} + \boldsymbol{\omega}_{2,1} + ... + \boldsymbol{\omega}_{n,n-1} = \mathbf{0} \quad \text{or} \quad \sum_{(i)} \boldsymbol{\omega}_{i,i-1} = \mathbf{0}. \tag{3.6}$$

The first vectorial equation for the angular velocities of a simple closed kinematic chain is given by Eq. (3.6).

The relation between the velocity $\mathbf{v}_{A_{i,i}}$, of the point $A_{i,i}$ on link (i), and the velocity $\mathbf{v}_{Ai,i-1}$, of the point $A_{i,i-1}$ on link $(i - 1)$, is

$$\mathbf{v}_{A_{i,i}} = \mathbf{v}_{A_{i,i-1}} + \mathbf{v}_{A_{i,i-1}}^r , \tag{3.7}$$

where $\mathbf{v}_{A_{i,i-1}}^r = \mathbf{v}_{A_{i,i} A_{i,i-1}}^r$ is the relative velocity of $A_{i,i}$ on link (i) with respect to $A_{i,i-1}$ on link $(i - 1)$, Fig. 3.2.

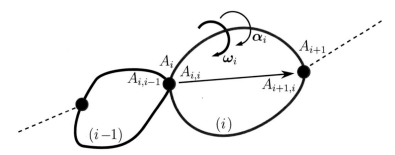

Fig. 3.2 Link $(i - 1)$ and link (i) connected at A_i

The velocities of two particles $A_{i,i}$ and $A_{i+1,i}$ on the same rigid body (i) are related by

$$\mathbf{v}_{A_{i+1,i}} = \mathbf{v}_{A_{i,i}} + \boldsymbol{\omega}_i \times \mathbf{r}_{A_i A_{i+1}} , \tag{3.8}$$

where $\boldsymbol{\omega}_i$ is the absolute angular velocity of the link (i) in the reference frame and $\mathbf{r}_{A_i A_{i+1}}$ is the distance vector from A_i to A_{i+1}.

Using Eqs. (3.7) and (3.8), the velocity of the point $A_{i+1,i} \in (i + 1)$ is written as

$$\mathbf{v}_{A_{i+1,i}} = \mathbf{v}_{A_{i,i-1}} + \boldsymbol{\omega}_i \times \mathbf{r}_{A_i A_{i+1}} + \mathbf{v}^r_{A_{i,i-1}} . \tag{3.9}$$

For the n link closed kinematic chain the following expressions are obtained

$$\mathbf{v}_{A_{3,2}} = \mathbf{v}_{A_{2,1}} + \boldsymbol{\omega}_2 \times \mathbf{r}_{A_2 A_3} + \mathbf{v}^r_{A_{2,1}}$$
$$\mathbf{v}_{A_{4,3}} = \mathbf{v}_{A_{3,2}} + \boldsymbol{\omega}_3 \times \mathbf{r}_{A_3 A_4} + \mathbf{v}^r_{A_{3,2}}$$
$$\dots\dots\dots\dots\dots\dots\dots\dots\dots\dots\dots\dots\dots\dots$$
$$\mathbf{v}_{A_{i+1,i}} = \mathbf{v}_{A_{i,i-1}} + \boldsymbol{\omega}_i \times \mathbf{r}_{A_i A_{i+1}} + \mathbf{v}^r_{A_{i,i-1}}$$
$$\dots\dots\dots\dots\dots\dots\dots\dots\dots\dots\dots\dots\dots\dots$$
$$\mathbf{v}_{A_{1,n}} = \mathbf{v}_{A_{n,n-1}} + \boldsymbol{\omega}_n \times \mathbf{r}_{A_n A_1} + \mathbf{v}^r_{A_{n,n-1}}$$
$$\mathbf{v}_{A_{2,1}} = \mathbf{v}_{A_{1,n}} + \boldsymbol{\omega}_1 \times \mathbf{r}_{A_1 A_2} + \mathbf{v}^r_{A_{1,n}} . \tag{3.10}$$

Summing the relations in Eq. (3.10):

$$\left(\boldsymbol{\omega}_1 \times \mathbf{r}_{A_1 A_2} + \boldsymbol{\omega}_2 \times \mathbf{r}_{A_2 A_3} + \dots + \boldsymbol{\omega}_i \times \mathbf{r}_{A_i A_{i+1}} + \dots + \boldsymbol{\omega}_n \times \mathbf{r}_{A_n A_1} \right)$$
$$+ \left(\mathbf{v}^r_{A_{2,1}} + \mathbf{v}^r_{A_{3,2}} + \dots + \mathbf{v}^r_{A_{i,i-1}} + \dots + \mathbf{v}^r_{A_{n,n-1}} + \mathbf{v}^r_{A_{1,n}} \right) = \mathbf{0}. \tag{3.11}$$

or

$$\left[\mathbf{r}_{A_1} \times (\boldsymbol{\omega}_1 - \boldsymbol{\omega}_n) + \mathbf{r}_{A_2} \times (\boldsymbol{\omega}_2 - \boldsymbol{\omega}_1) + \dots + \mathbf{r}_{A_n} \times (\boldsymbol{\omega}_n - \boldsymbol{\omega}_{n-1}) \right]$$
$$+ \left(\mathbf{v}^r_{A_{1,n}} + \mathbf{v}^r_{A_{2,1}} + \dots + \mathbf{v}^r_{A_{i,i-1}} + \dots + \mathbf{v}^r_{A_{n,n-1}} \right) = \mathbf{0},$$

$$\left(\mathbf{r}_{A_1} \times \boldsymbol{\omega}_{1,n} + \mathbf{r}_{A_2} \times \boldsymbol{\omega}_{2,1} + \ldots + \mathbf{r}_{A_n} \times \boldsymbol{\omega}_{n,n-1} \right)$$
$$+ \left(\mathbf{v}^r_{A_{1,n}} + \mathbf{v}^r_{A_{2,1}} + \ldots + \mathbf{v}^r_{A_{n,n-1}} \right) = \mathbf{0}. \tag{3.12}$$

Equation (3.12) represents the second vectorial equation for the angular velocities of a simple closed kinematic chain.

The velocity equations for a simple closed kinematic chain are [3, 71]

$$\sum_{(i)} \boldsymbol{\omega}_{i,i-1} = \mathbf{0} \ \text{and} \ \sum_{(i)} \mathbf{r}_{A_i} \times \boldsymbol{\omega}_{i,i-1} + \sum_{(i)} \mathbf{v}^r_{A_{i,i-1}} = \mathbf{0}. \tag{3.13}$$

The vectorial acceleration equations for simple closed kinematic chain are [3, 71]

$$\sum_{(i)} \boldsymbol{\alpha}_{i,i-1} + \sum_{(i)} \boldsymbol{\omega}_i \times \boldsymbol{\omega}_{i,i-1} = \mathbf{0} \ \text{and}$$

$$\sum_{(i)} \mathbf{r}_{A_i} \times (\boldsymbol{\alpha}_{i,i-1} + \boldsymbol{\omega}_i \times \boldsymbol{\omega}_{i,i-1}) + \sum_{(i)} \mathbf{a}^r_{A_{i,i-1}} + \sum_{(i)} \mathbf{a}^c_{A_{i,i-1}}$$

$$+ \sum_{(i)} \boldsymbol{\omega}_i \times (\boldsymbol{\omega}_i \times \mathbf{r}_{A_i A_{i+1}}) = \mathbf{0}. \tag{3.14}$$

The relative Coriolis acceleration is

$$\mathbf{a}^c_{A_{i,i-1}} = 2\boldsymbol{\omega}_{i-1} \times \mathbf{v}^r_{A_{i,i-1}}. \tag{3.15}$$

The equations for velocities and accelerations are written for any closed loop of the mechanism. It is best to write the closed loops equations only for the independent loops of the diagram representing the mechanism.

For a closed kinematic chain in planar motion the following equations will be used:

$$\sum_{(i)} \boldsymbol{\omega}_{i,i-1} = \mathbf{0},$$

$$\sum_{(i)} \mathbf{r}_{A_i} \times \boldsymbol{\omega}_{i,i-1} + \sum_{(i)} \mathbf{v}^r_{A_{i,i-1}} = \mathbf{0}. \tag{3.16}$$

$$\sum_{(i)} \boldsymbol{\alpha}_{i,i-1} = \mathbf{0},$$

$$\sum_{(i)} \mathbf{r}_{A_i} \times \boldsymbol{\alpha}_{i,i-1} + \sum_{(i)} \mathbf{a}^r_{A_{i,i-1}} + \sum_{(i)} \mathbf{a}^c_{A_{i,i-1}} - \omega_i^2 \mathbf{r}_{A_i A_{i+1}} = \mathbf{0}. \tag{3.17}$$

Independent closed contours method can be summarized as:

Step 1. Determine the position analysis of the mechanism.

Step 2. Draw a diagram representing the mechanism and select the independent closed loops. Determine a path for each closed loop.

Step 3. For each closed loop write the closed loop velocity relations and the closed loops acceleration relations.

Step 4. Solve the linear algebraic equations with the unknowns as the components of the relative angular velocities $\omega_{j,j-1}$; the components of the relative angular accelerations $\alpha_{j,j-1}$; the components of the relative linear velocities $v^r_{Aj,j-1}$; the components of the relative linear accelerations $a^r_{Aj,j-1}$. Solve the algebraic system of equations and determine the unknown kinematic parameters.

Step 5. Determine the absolute angular velocities ω_j and the absolute angular accelerations α_j. Find the absolute linear velocities and accelerations of points and joints.

3.2 Closed Contour Equations for R-RTR Mechanism

The planar R-RTR mechanism considered in Chap. 2 is shown in Fig. 3.3 with the attached contour diagram. The input data and the position data are:

```
AB = 0.10 ;   % (m)
AC = 0.05 ;   % (m)
CD = 0.15 ;   % (m)
phi = pi/6;   % (rad)

xA = 0; yA = 0;
rA_ = [xA yA 0];
xC = AC ; yC = 0;
rC_ = [xC yC 0];
```

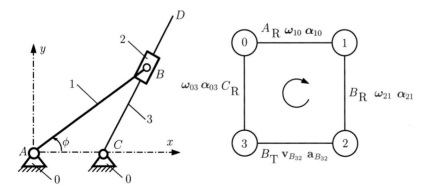

Fig. 3.3 R-RTR mechanism and contour diagram

```
% rC_  = [ 0.050, 0.000,0] (m)
% rB_  = [ 0.087, 0.050,0] (m)
% phi2 = phi3 = 53.794 (degrees)
% rD_  = [ 0.139, 0.121,0] (m)
```

The mechanism has one contour. The contour is made of 0, 1, 2, 3, and 0. A clockwise path is chosen for the contour. The contour has: a rotational joint R between the links 0 and 1 (joint A_R); a rotational joint R between the links 1 and 2 (joint B_R); a translational joint T between the links 2 and 3 (joint B_T); a rotational joint R between the links 3 and 0 (joint C_R). The angular velocity ω_{10} of the driver link is $\omega_{10} = \omega_1 = \omega = n\,\pi/30 = -50\pi/30$ rad/s $= -5\,\pi/3$ rad/s. The velocity equations for the contour are:

$$\boldsymbol{\omega}_{10} + \boldsymbol{\omega}_{21} + \boldsymbol{\omega}_{03} = \mathbf{0},$$
$$\mathbf{r}_B \times \boldsymbol{\omega}_{21} + \mathbf{r}_C \times \boldsymbol{\omega}_{03} + \mathbf{v}^{\text{rel}}_{B_3 B_2} = \mathbf{0}, \tag{3.18}$$

where $\mathbf{r}_B = x_B\,\mathbf{\imath} + y_B\,\mathbf{J}$, $\mathbf{r}_C = x_C\,\mathbf{\imath} + y_C\,\mathbf{J}$, and

$$\boldsymbol{\omega}_{10} = \omega_{10}\,\mathbf{k}, \quad \boldsymbol{\omega}_{21} = \omega_{21}\,\mathbf{k}, \quad \boldsymbol{\omega}_{03} = \omega_{03}\,\mathbf{k},$$
$$\mathbf{v}^{\text{rel}}_{B_3 B_2} = \mathbf{v}_{B_{32}} = v_{B_{32}}\cos\phi_2\,\mathbf{\imath} + v_{B_{32}}\sin\phi_2\,\mathbf{J}.$$

The relative angular velocities ω_{21}, ω_{03}, and relative linear velocity $v_{B_{32}}$ are the unknowns. Equation (3.18) can be written as

$$\omega_{10}\,\mathbf{k} + \omega_{21}\,\mathbf{k} + \omega_{03}\,\mathbf{k} = \mathbf{0},$$

$$\begin{bmatrix} \mathbf{\imath} & \mathbf{J} & \mathbf{k} \\ x_B & y_B & 0 \\ 0 & 0 & \omega_{21} \end{bmatrix} + \begin{bmatrix} \mathbf{\imath} & \mathbf{J} & \mathbf{k} \\ x_C & y_C & 0 \\ 0 & 0 & \omega_{03} \end{bmatrix} + v_{B_{32}}\cos\phi_2\,\mathbf{\imath} + v_{B_{32}}\sin\phi_2\,\mathbf{J} = \mathbf{0}. \tag{3.19}$$

Equation (3.19) represents a system of 3 algebraic equations with 3 unknowns and can be written in MATLAB as:

```
n = -50.; % (rpm)
omega1_ = [0 0 pi*n/30]; % (rad/s)

% symbolic unknowns
syms omega21z omega03z vB32

omega10_ = omega1_;
omega21v_ = [ 0 0 omega21z];
omega03v_ = [ 0 0 omega03z];
v32v_ = vB32*[ cos(phi2) sin(phi2) 0];

eqIomega_ = omega10_ + omega21v_ + omega03v_;
```

```
eqIvz = eqIomega_(3);
eqIv = cross(rB_,omega21v_) + cross(rC_,omega03v_) + v32v_;
eqIvx=eqIv(1);
eqIvy=eqIv(2);
```

The system of algebraic equations is solved in MATLAB:

```
solIv=solve(eqIvz,eqIvx,eqIvy);
omega21_ = [ 0 0 eval(solIv.omega21z) ];
omega03_ = [ 0 0 eval(solIv.omega03z) ];
vB3B2_ = eval(solIv.vB32)*[ cos(phi2) sin(phi2) 0];

fprintf('omega21_ = [ %g, %g, %6.3f ] (rad/s)\n', omega21_)
fprintf('omega03_ = [ %g, %g, %6.3f ] (rad/s)\n', omega03_)
fprintf('vB32 = %6.3f (m/s)\n', eval(solIv.vB32))
fprintf('vB3B2_ = [ %6.3f, %6.3f, %d ] (m/s)\n', vB3B2_)
```

The angular velocity of link 3 is

$$\omega_{30} = -\omega_{03},$$

the angular velocity of the link 2 is $\omega_{20} = \omega_{30}$, and the velocity of point D is

$$\mathbf{v}_D = \mathbf{v}_C + \omega_{30} \times (\mathbf{r}_D - \mathbf{r}_C),$$

and the MATLAB code is:

```
omega30_ = - omega03_;
omega20_ = omega30_;
vC_ = [0 0 0]; % C is fixed
vD_ = vC_ + cross(omega30_, rD_-rC_);
fprintf('omega20_=omega30_= [%d, %d, %6.3f] (rad/s)\n',omega30_)
fprintf('vD_ = [ %6.3f, %6.3f, %g ] (m/s)\n', vD_)

% vB32 = 0.211 (m/s)
% vB3B2_ = [ 0.125, 0.170, 0 ] (m/s)
% omega20_=omega30_= [0, 0, -7.732] (rad/s)
% vD_ = [ 0.936, -0.685, 0 ] (m/s)
```

The closed contour equations for the acceleration field are

$$\boldsymbol{\alpha}_{10} + \boldsymbol{\alpha}_{21} + \boldsymbol{\alpha}_{03} = \mathbf{0},$$
$$\mathbf{r}_B \times \boldsymbol{\alpha}_{21} + \mathbf{r}_C \times \boldsymbol{\alpha}_{03} + \mathbf{a}_{B_3 B_2}^{\text{rel}} + \mathbf{a}_{B_3 B_2}^{\text{cor}} - \omega_{10}^2(\mathbf{r}_B - \mathbf{r}_A) - \omega_{20}^2(\mathbf{r}_C - \mathbf{r}_B) = \mathbf{0}. \quad (3.20)$$

The angular accelerations are

$$\boldsymbol{\alpha}_{10} = \alpha_{10}\,\mathbf{k}, \quad \boldsymbol{\alpha}_{21} = \alpha_{21}\,\mathbf{k}, \quad \boldsymbol{\alpha}_{03} = \alpha_{03}\,\mathbf{k}.$$

The linear relative accelerations of B_3 about 2 is

$$\mathbf{a}^{rel}_{B_3 B_2} = \mathbf{a}_{B_{32}} = a_{B_{32}} \cos \phi_2 \, \mathbf{I} + a_{B_{32}} \sin \phi_2 \, \mathbf{J}.$$

The Coriolis relative acceleration of B_3 about 2 is

$$\mathbf{a}^{cor}_{B_3 B_2} = \mathbf{a}^{c}_{B_{32}} = 2 \, \boldsymbol{\omega}_{20} \times \mathbf{v}_{B_{32}} = 2 \, \boldsymbol{\omega}_{30} \times \mathbf{v}_{B_{32}}.$$

The angular acceleration of the driver link is $\alpha_{10} = \dot{\omega}_{10} = 0$. In MATLAB the unknown relative accelerations are introduced as symbolical variables:

```
syms alpha21z alpha03z aB32
alpha10_ = [ 0 0 0 ];
alpha21v_ = [ 0 0 alpha21z];
alpha03v_ = [ 0 0 alpha03z];
a32v_ = aB32*[ cos(phi2) sin(phi2) 0];
```

and the system of equations can be written as:

```
eqIalpha_ = alpha10_ + alpha21v_ + alpha03v_;
eqIaz = eqIalpha_(3);
eqIa_=cross(rB_,alpha21v_)+cross(rC_,alpha03v_)...
      +a32v_+2*cross(omega20_,vB3B2_)...
-dot(omega1_,omega1_)*rB_-dot(omega20_,omega20_)*(rC_-rB_);
eqIax = eqIa_(1);
eqIay = eqIa_(2);
```

The system of three algebraic equations is solved with MATLAB:

```
solIa = solve(eqIaz,eqIax,eqIay);
alpha21_ = [ 0 0 eval(solIa.alpha21z) ];
alpha03_ = [ 0 0 eval(solIa.alpha03z) ];
aB3B2_ = eval(solIa.aB32)*[ cos(phi2) sin(phi2) 0];

fprintf('alpha21_ = [ %g, %g, %6.3f ] (rad/s^2)\n', alpha21_)
fprintf('alpha03_ = [ %g, %g, %6.3f ] (rad/s^2)\n', alpha03_)
fprintf('aB32 = %6.3f (m/s^2)\n', eval(solIa.aB32))
fprintf('aB3B2_ = [ %6.3f, %6.3f, %d ] (m/s^2)\n', aB3B2_)
%alpha21_ = [ 0, 0, -34.865 ] (rad/s^2)
%alpha03_ = [ 0, 0, 34.865 ] (rad/s^2)
%aB32 = -1.196 (m/s^2)
%aB3B2_ = [ -0.706, -0.965, 0 ] (m/s^2)
```

The absolute accelerations are:

```
alpha30_ = - alpha03_;
alpha20_ = alpha30_;
aC_ = [0 0 0 ];
aD_=aC_+cross(alpha30_,rD_-rC_)-dot(omega20_,omega20_)*(rD_-rC_);
fprintf('alpha20_=alpha30_=[%d,%d,%6.3f] (rad/s^2)\n',alpha30_)
fprintf('aD_ = [ %6.3f, %6.3f, %g ] (m/s^2)\n', aD_)
% alpha20_=alpha30_=[0,0,-34.865] (rad/s^2)
% aD_ = [ -1.077, -10.324, 0 ] (m/s^2)
```

The results obtained with the closed contour method are identical with the results obtained with the classical, and derivative methods.

3.3 Force Analysis for R-RTR Mechanism

The inertia forces and moments and the gravity forces on links 2 and 3 are shown in Fig. 3.4a. The input data for joint force analysis are:

```
% rC_ = [ 0.050, 0.000,0] (m)
% rB_ = [ 0.087, 0.050,0] (m)
% phi2 = phi3 = 53.794 (degrees)
% rD_ = [ 0.139, 0.121,0] (m)
% rC1_ = [ 0.043, 0.025,0] (m)
% rC2_ = [ 0.087, 0.050,0] (m)
% rC3_ = [ 0.094, 0.061,0] (m)
% aC1_ = [-1.187,-0.685,0] (m/s^2)
% aC2_ = [-2.374,-1.371,0] (m/s^2)
% aC3_ = [-0.538,-5.162,0] (m/s^2)
% alpha1_ = [0 ,0, 0.000](rad/s^2)
% alpha2_ = [0,0,-34.865](rad/s^2)
% alpha3_ = [0,0,-34.865](rad/s^2)
%
% G1_ = - m1 g_ = [0,-0.078,0] (N)
% Fin1_= - m1 aC1_ = [ 0.009, 0.005,0] (N)
% Min1_=-IC1 alpha1_=[0, 0, 0](N m)
%
% G2_ = - m2 g_ = [0,-0.063,0] (N)
% Fin2_= - m2 aC2_ =[ 0.015, 0.009,0] (N)
% Min2_=-IC2 alpha2_=[0,0,3.719e-05](N m)
%
% G3_ = - m3 g_ = [0, -0.118, 0] (N)
% Fin3_ = - m3 aC3_ =[ 0.006, 0.062,0] (N)
% Min3_=-IC3 alpha3_=[0,0,7.880e-04](N m)
% Me_ = [0, 0, 100.000] (N m)
```

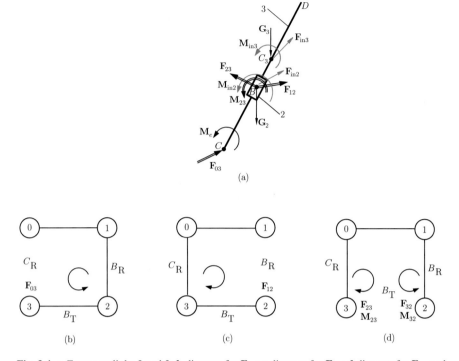

Fig. 3.4 **a** Forces on links 2 and 3; **b** diagram for F_{03}; **c** diagram for F_{12}; **d** diagram for F_{23} and M_{23}

To calculate the joint reaction $\mathbf{F}_{03}$ with the application point at the rotational joint C the diagram shown in Fig. 3.4(b) is used. For the link 3, the joint reaction force $\mathbf{F}_{23}$ is eliminated if the sum of the forces on link 3 are dot multiplied with the direction of the link BC:

```
syms F03x F03y
F03_=[F03x, F03y, 0]; % unknown joint force of 0 on 3 at C
% for joint B translation
% (F03_+G3_+Fin3_+F23_) .(rB_-rC_)=0
% F23_ perpendicular to BC: F32_.BC_ = 0 =>
eqF3 = (F03_+G3_+Fin3_)*(rB_-rC_).'; % (F1)
```

Another equations for the unknown $\mathbf{F}_{03}$ is obtained taking sum of the moments for links 3 and 2 about the rotational joint at $B = C_2$:

```
% for joint B rotation
% BC_xF03_+BC3_x(G3_+Fin3_)+Me_+Min2_+Min3_=0_
eqM32B_ = ...
cross(rC_-rB_,F03_)+cross(rC3_-rB_,G3_+Fin3_)...
```

```
+Me_+Min2_+Min3_;
eqM32z = eqM32B_(3); % (F2)
```

Solving just two algebraic equations the force $\mathbf{F}_{03}$ is calculated:

```
% Eq(F1) and Eq(F2) =>
solF03=solve(eqF3,eqM32z);
F03xs=eval(solF03.F03x);
F03ys=eval(solF03.F03y);
F03s_=[F03xs, F03ys, 0];
fprintf('F03_ = [%6.3f,%6.3f,%g] (N)\n',F03s_)
% F03_ = [-1302.149,953.291,0] (N)
```

The unknown joint reaction force at B rotation is the force of link 1 on link 2, $\mathbf{F}_{12}$. From the diagram on Fig. 3.4c the following equations are developed. For the joint B translation, the sum of the forces on link 2 dot multiplied with the direction of the link BC is considered:

```
syms F12x F12y
F12_=[F12x, F12y, 0]; % unknown joint force of 1 on 2 at B
% for joint B translation
% (F12_+G2_+Fin2_+F32_) .(rB_-rC_)=0
% F32_ = -F23_ perpendicular to BC: F32_.BC_ = 0 =>
eqF2 = (F12_+G2_+Fin2_)*(rB_-rC_).'; % (F3)
```

For the joint C rotation, the sum of the moments on links 2 and 3 about C is:

```
% for joint C rotation
% CB_x(F12_+G2_+Fin2_)+CC3_x(G3_+Fin3_)+Me_+Min2_+Min3_=0_
eqM23C_ = ...
cross(rB_-rC_,F12_+G2_+Fin2_)+cross(rC3_-rC_,G3_+Fin3_)...
+Me_+Min2_+Min3_;
eqM23z = eqM23C_(3); % (F4)
```

There are two equations and the force $\mathbf{F}_{12}$ is calculated with:

```
% Eq(F3) and Eq(F4) =>
solF12=solve(eqF2,eqM23z);
F12xs=eval(solF12.F12x);
F12ys=eval(solF12.F12y);
F12s_=[F12xs, F12ys, 0];
fprintf('F12_ = [%6.3f,%6.3f,%g] (N)\n',F12s_)
% F12_ = [1302.128,-953.181,0] (N)
```

The joint reaction at B translation is replaced by a force $\mathbf{F}_{23}$ at B and a joint reaction moment $\mathbf{M}_{23}$, as shown in Fig. 3.4d. To calculate the joint reaction moment $\mathbf{M}_{23}$ the sum of the moments on slider 2 about B rotation gives:

```
% for joint B rotation
% Min2_ + M32_ = 0_  => Min2_ - M23_ = 0_
M23_= Min2_; %
fprintf('M23_ = [%d,%d,%6.6f] (N m)\n',M23_)
% M23_ = [0,0,0.000037] (N m)
```

The forces on link 2 give a zero moment about B. The joint reaction force $\mathbf{F}_{23}$ at B is perpendicular to BC and the the sum of the moments on link 3 about C rotation give:

```
syms F23x F23y
F23_=[F23x, F23y, 0]; % unknown joint force of 2 on 3 at B
% F23_ perpendicular to BC: F32.BC = 0 =>
eqF23 = (rB_-rC_)*F23_.'; % (F5)
% for joint C rotation
% CB_xF23_+CC3_x(G3_+Fin3_)+Me_+Min3_+Min3_=0_
eqM3C_ = ...
cross(rB_-rC_,F23_)+cross(rC3_-rC_,G3_+Fin3_)...
+Me_+Min3_+M23_;
eqM3Cz = eqM3C_(3); % (F6)
```

The force $\mathbf{F}_{23}$ is:

```
% Eq(F5) and Eq(F6) =>
solF23=solve(eqF23,eqM3Cz);
F23xs=eval(solF23.F23x);
F23ys=eval(solF23.F23y);
F23s_=[F23xs, F23ys, 0];
fprintf('F23_ = [%6.3f,%6.3f,%g] (N)\n',F23s_)
% F23_ = [1302.143,-953.235,0] (N)
```

Figure 3.5 shows the diagram for calculating the joint reaction force of the ground on driver 1 at A, $\mathbf{F}_{01}$, and the equilibrium moment (motor moment) $\mathbf{M}_m$. The first algebraic equation is the sum of the moments for driver 1 about B (Fig. 3.5):

```
syms F01x F01y Mm
F01_=[F01x, F01y, 0]; % unknown joint force of 0 on 1 at A
Mm_= [0, 0, Mm]; % equilibrium moment
% for joint B rotation
% BA_xF01_+BC1_x(G1_+Fin1_)+Mm_+Min1_=0_
eqM1B_ = ...
```

Fig. 3.5 Diagram for $\mathbf{F}_{01}$
and $\mathbf{M}_m$

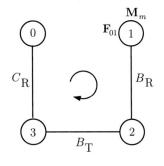

```
 cross(rA_-rB_,F01_)+cross(rC1_-rB_,G1_+Fin1_)+Mm_+Min1_;
 eqM1Bz = eqM1B_(3); % (F7)
```

The second equation is the sum of the forces on links 1 and 2 projected on the *BC* direction:

```
% for joint B translation
% (F01+G1_+Fin1_+G2_+Fin2_+F32_) .(rB_-rC_)=0
% F32_ = -F23_ perpendicular to BC: F32_.BC_ = 0 =>
 eqF12 = (F01_+G1_+Fin1_+G2_+Fin2_)*(rB_-rC_).'; % (F8)
```

The last equations is the sum of the moments of links 1, 2, and 3 about *C*:

```
% for joint C rotation
% CA_xF01_+CC1_x(G1_+Fin1_)+CB_x(G2_+Fin2_)+CC3_x(G3_+Fin3_)
% +Mm_+Me_+Min2_+Min3_=0_
 eqM123C_ = ...
  cross(rA_-rC_,F01_)+cross(rC1_-rC_,G1_+Fin1_)...
 +cross(rB_-rC_,G2_+Fin2_)+cross(rC3_-rC_,G3_+Fin3_)...
 +Mm_+Me_+Min2_+Min3_;
 eqM123z = eqM123C_(3); % (F9)
```

There are three equations with three unknowns, F_{01x}, F_{01y}, and M_m:

```
% Eq(F7), Eq(F8) and Eq(F9) =>
solF01=solve(eqM1Bz, eqF12, eqM123z);
F01xs=eval(solF01.F01x);
F01ys=eval(solF01.F01y);
Mms=eval(solF01.Mm);
F01s_=[F01xs, F01ys, 0];
Mms_=[0, 0, Mms];
fprintf('F01_ = [%6.3f,%6.3f,%g] (N)\n',F01s_)
fprintf('Mm_ = [%d,%d,%6.3f] (N m)\n',Mms_)
```

```
% F01_ = [1302.118,-953.108,0] (N)
% Mm_ = [0,0,-147.651] (N m)
```

3.4 Problems

3.1 Calculate the kinematics and the dynamics of the system in Problem 2.1 using the contour methods. Check the results with the two methods.

3.2 For Problem 2.2 find the velocities, accelerations, and joint forces with the contour methods. Check the results with the two methods.

3.3 With the help of contour methods and using the input data given in Problem 2.3 determine the velocities, accelerations, and joint forces for the mechanism.

References

1. E.A. Avallone, T. Baumeister, A. Sadegh, *Marks' Standard Handbook for Mechanical Engineers*, 11th edn. (McGraw-Hill Education, New York, 2007)
2. I.I. Artobolevski, *Mechanisms in Modern Engineering Design* (MIR, Moscow, 1977)
3. M. Atanasiu, *Mechanics* (Mecanica) (EDP, Bucharest, 1973)
4. H. Baruh, *Analytical Dynamics* (WCB/McGraw-Hill, Boston, 1999)
5. A. Bedford, W. Fowler, *Dynamics* (Addison Wesley, Menlo Park, CA, 1999)
6. A. Bedford, W. Fowler, *Statics* (Addison Wesley, Menlo Park, CA, 1999)
7. F.P. Beer et al., *Vector Mechanics for Engineers: Statics and Dynamics* (McGraw-Hill, New York, NY, 2016)
8. M. Buculei, D. Bagnaru, G. Nanu, D.B. Marghitu, *Computing Methods in the Analysis of the Mechanisms with Bars* (Scrisul Romanesc, Craiova, 1986)
9. M.I. Buculei, *Mechanisms* (University of Craiova Press, Craiova, Romania, 1976)
10. D. Bolcu, S. Rizescu, *Mecanica* (EDP, Bucharest, 2001)
11. J. Billingsley, *Essential of Dynamics and Vibration* (Springer, 2018)
12. R. Budynas, K.J. Nisbett, *Shigley's Mechanical Engineering Design*, 9th edn. (McGraw-Hill, New York, 2013)
13. J.A. Collins, H.R. Busby, G.H. Staab, *Mechanical Design of Machine Elements and Machines*, 2nd edn. (Wiley, 2009)
14. M. Crespo da Silva, *Intermediate Dynamics for Engineers* (McGraw-Hill, New York, 2004)
15. A.G. Erdman, G.N. Sandor, *Mechanisms Design* (Prentice-Hall, Upper Saddle River, NJ, 1984)
16. M. Dupac, D.B. Marghitu, *Engineering Applications: Analytical and Numerical Calculation with MATLAB* (Wiley, Hoboken, NJ, 2021)
17. A. Ertas, J.C. Jones, *The Engineering Design Process* (Wiley, New York, 1996)
18. F. Freudenstein, An application of Boolean Algebra to the Motion of Epicyclic Drivers. J. Eng. Ind. 176–182 (1971)
19. J.H. Ginsberg, *Advanced Engineering Dynamics* (Cambridge University Press, Cambridge, 1995)
20. H. Goldstein, *Classical Mechanics* (Addison-Wesley, Redwood City, CA, 1989)
21. D.T. Greenwood, *Principles of Dynamics* (Prentice-Hall, Englewood Cliffs, NJ, 1998)
22. A.S. Hall, A.R. Holowenko, H.G. Laughlin, *Schaum's Outline of Machine Design* (McGraw-Hill, New York, 2013)

23. B.G. Hamrock, B. Jacobson, S.R. Schmid, *Fundamentals of Machine Elements* (McGraw-Hill, New York, 1999)
24. R.C. Hibbeler, *Engineering Mechanics: Dynamics* (Prentice Hall, 2010)
25. T.E. Honein, O.M. O'Reilly, On the Gibbs–Appell equations for the dynamics of rigid bodies. J. Appl. Mech. **88**/074501-1 (2021)
26. R.C. Juvinall, K.M. Marshek, *Fundamentals of Machine Component Design*, 5th edn. (Wiley, New York, 2010)
27. T.R. Kane, *Analytical Elements of Mechanics*, vol. 1 (Academic Press, New York, 1959)
28. T.R. Kane, *Analytical Elements of Mechanics*, vol. 2 (Academic Press, New York, 1961)
29. T.R. Kane, D.A. Levinson, The use of Kane's dynamical equations in robotics. MIT Int. J. Robot. Res. **3**, 3–21 (1983)
30. T.R. Kane, P.W. Likins, D.A. Levinson, *Spacecraft Dynamics* (McGraw-Hill, New York, 1983)
31. T.R. Kane, D.A. Levinson, *Dynamics* (McGraw-Hill, New York, 1985)
32. K. Lingaiah, *Machine Design Databook*, 2nd edn. (McGraw-Hill Education, New York, 2003)
33. N.I. Manolescu, F. Kovacs, A. Oranescu, *The Theory of Mechanisms and Machines* (EDP, Bucharest, 1972)
34. D.B. Marghitu, *Mechanical Engineer's Handbook* (Academic Press, San Diego, CA, 2001)
35. D.B. Marghitu, M.J. Crocker, *Analytical Elements of Mechanisms* (Cambridge University Press, Cambridge, 2001)
36. D.B. Marghitu, *Kinematic Chains and Machine Component Design* (Elsevier, Amsterdam, 2005)
37. D.B. Marghitu, *Mechanisms and Robots Analysis with MATLAB* (Springer, New York, NY, 2009)
38. D.B. Marghitu, M. Dupac, *Advanced Dynamics: Analytical and Numerical Calculations with MATLAB* (Springer, New York, NY, 2012)
39. D.B. Marghitu, M. Dupac, H.M. Nels, *Statics with MATLAB* (Springer, New York, NY, 2013)
40. D.B. Marghitu, D. Cojocaru, *Advances in Robot Design and Intelligent Control* (Springer International Publishing, Cham, Switzerland, 2016), pp. 317–325
41. D.J. McGill, W.W. King, *Engineering Mechanics: Statics and an Introduction to Dynamics* (PWS Publishing Company, Boston, 1995)
42. J.L. Meriam, L.G. Kraige, *Engineering Mechanics: Dynamics* (Wiley, New York, 2007)
43. C.R. Mischke, Prediction of Stochastic Endurance strength. Trans. ASME J. Vib. Acoust. Stress Reliab. Des. **109**(1), 113–122 (1987)
44. L. Meirovitch, *Methods of Analytical Dynamics* (Dover, 2003)
45. R.L. Mott, *Machine Elements in Mechanical Design* (Prentice Hall, Upper Saddle River, NJ, 1999)
46. W.A. Nash, *Strength of Materials* (Schaum's Outline Series, McGraw-Hill, New York, 1972)
47. R.L. Norton, *Machine Design* (Prentice-Hall, Upper Saddle River, NJ, 1996)
48. R.L. Norton, *Design of Machinery* (McGraw-Hill, New York, 1999)
49. O.M. O'Reilly, *Intermediate Dynamics for Engineers Newton-Euler and Lagrangian Mechanics* (Cambridge University Press, UK, 2020)
50. O.M. O'Reilly, *Engineering Dynamics: A Primer* (Springer, NY, 2010)
51. W.C. Orthwein, *Machine Component Design* (West Publishing Company, St. Paul, 1990)
52. L.A. Pars, *A Treatise on Analytical Dynamics* (Wiley, New York, 1965)
53. F. Reuleaux, *The Kinematics of Machinery* (Dover, New York, 1963)
54. D. Planchard, M. Planchard, *SolidWorks 2013 Tutorial with Video Instruction* (SDC Publications, 2013)
55. I. Popescu, *Mechanisms* (University of Craiova Press, Craiova, Romania, 1990)
56. I. Popescu, C. Ungureanu, *Structural Synthesis and Kinematics of Mechanisms with Bars* (Universitaria Press, Craiova, Romania, 2000)
57. I. Popescu, L. Luca, M. Cherciu, D.B. Marghitu, *Mechanisms for Generating Mathematical Curves* (Springer Nature, Switzerland, 2020)
58. C.A. Rubin, *The Student Edition of Working Model* (Addison-Wesley Publishing Company, Reading, MA, 1995)

59. J. Ragan, D.B. Marghitu, Impact of a Kinematic link with MATLAB and SolidWorks. Appl. Mech. Mater. **430**, 170–177 (2013)
60. J. Ragan, D.B. Marghitu, MATLAB dynamics of a free link with elastic impact, *International Conference on Mechanical Engineering, ICOME 2013*, 16–17 May 2013, Craiova, Romania, 2013
61. J.C. Samin, P. Fisette, *Symbolic Modeling of Multibody Systems* (Kluwer, 2003)
62. A.A. Shabana, *Computational Dynamics* (Wiley, New York, 2010)
63. I.H. Shames, *Engineering Mechanics—Statics and Dynamics* (Prentice-Hall, Upper Saddle River, NJ, 1997)
64. J.E. Shigley, C.R. Mischke, *Mechanical Engineering Design* (McGraw-Hill, New York, 1989)
65. J.E. Shigley, C.R. Mischke, R.G. Budynas, *Mechanical Engineering Design*, 7th edn. (McGraw-Hill, New York, 2004)
66. J.E. Shigley, J.J. Uicker, *Theory of Machines and Mechanisms* (McGraw-Hill, New York, 1995)
67. D. Smith, *Engineering Computation with MATLAB* (Pearson Education, Upper Saddle River, NJ, 2008)
68. R.W. Soutas-Little, D.J. Inman, *Engineering Mechanics: Statics and Dynamics* (Prentice-Hall, Upper Saddle River, NJ, 1999)
69. J. Sticklen, M.T. Eskil, *An Introduction to Technical Problem Solving with MATLAB* (Great Lakes Press, Wildwood, MO, 2006)
70. A.C. Ugural, *Mechanical Design* (McGraw-Hill, New York, 2004)
71. R. Voinea, D. Voiculescu, V. Ceausu, *Mechanics* (Mecanica), EDP, Bucharest, 1983)
72. K.J. Waldron, G.L. Kinzel, *Kinematics, Dynamics, and Design of Machinery* (Wiley, New York, 1999)
73. J.H. Williams Jr., *Fundamentals of Applied Dynamics* (Wiley, New York, 1996)
74. C.E. Wilson, J.P. Sadler, *Kinematics and Dynamics of Machinery* (Harper Collins College Publishers, New York, 1991)
75. H.B. Wilson, L.H. Turcotte, and D. Halpern, *Advanced Mathematics and Mechanics Applications Using MATLAB* (Chapman & Hall/CRC, 2003)
76. S. Wolfram, *Mathematica* (Wolfram Media/Cambridge University Press, Cambridge, 1999)
77. National Council of Examiners for Engineering and Surveying (NCEES), *Fundamentals of Engineering. Supplied-Reference Handbook*, Clemson, SC, 2001
78. * * *, *The Theory of Mechanisms and Machines* (Teoria mehanizmov i masin) (Vassaia scola, Minsk, Russia, 1970)
79. eCourses—University of Oklahoma. http://ecourses.ou.edu/home.htm
80. https://www.mathworks.com
81. http://www.eng.auburn.edu/~marghitu/
82. https://www.solidworks.com
83. https://www.wolfram.com

Chapter 4
Classical Analysis of a Mechanism with Two Dyads

Abstract A planar mechanism with a driver link and two dyads is analyzed using vectorial equations for velocities and accelerations. Newton-Euler equations of motions are developed for the force analysis. Symbolical MATLAB is applied for the manipulation of the equations.

The planar R-RRT-RTR mechanism is shown in Fig. 4.1. The driver link is the rigid link 1 (the link AB). The first dyad is RRT formed by the elements 2 and 3, (B_R, D_R, D_T), and the second dyad is RTR composed by links 4 and 5, (C_R, E_T, E_R). The mechanism has five moving links. The following numerical data are given: $AB = 0.20$ m, $BC = 0.21$ m, $CD = 0.39$ m, $CF = 0.45$ m, $a = 0.30$ m, $b = 0.25$ m, and $c = 0.05$ m. The constant angular speed of the driver link 1 is $n = -500$ rpm. Find the velocity and the acceleration fields of the mechanism when the angle of the driver link 1 with the horizontal axis is $\phi = 45°$.

The external moment applied on the last link 5 is opposed to the motion of the link and has the value $M_{5e} = 1000$ N m. Find the motor moment M_m required for the dynamic equilibrium of the mechanism when the the driver link 1 makes an angle $\phi = 45°$ with the horizontal axis.

All the links are homogeneous rectangular prisms with the mass density $\rho = 8000$ kg/m³. The links 1, 2, and 4 are considered slender bars with the height $h = 0.02$ m and the depth $d = 0.01$ m. The sliders 3 and 5 have the height $h_S = 0.04$ m, the width $w_S = 0.08$ m, and the depth $d_S = 0.02$ m. The gravitational acceleration is $g = 9.807$ m/s².

4.1 Position Analysis

The input data are:

```
AB = 0.20 ; % m
BC = 0.21 ; % m
CD = 0.39 ; % m
```

© The Author(s), under exclusive license to Springer Nature Switzerland AG 2022
D. B. Marghitu et al., *Mechanical Simulation with MATLAB®*,
Springer Tracts in Mechanical Engineering,
https://doi.org/10.1007/978-3-030-88102-3_4

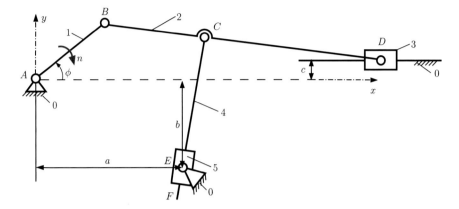

Fig. 4.1 R-RRT-RTR mechanism

```
CF  = 0.45 ;  % m
a   = 0.30 ;  % m
b   = 0.25 ;  % m
c   = 0.05 ;  % m
phi = 45*pi/180 ;  % rad
n = -500 ;          % rpm
```

A Cartesian reference frame xy is selected. The joint A is in the origin of the reference frame, that is, $A \equiv O$,

$$x_A = 0 \quad \text{and} \quad y_A = 0.$$

The coordinates of the joint E are

$$x_E = a \quad \text{and} \quad y_E = -b.$$

Position of joint B
The coordinates of the joint B are computed as following expressions

$$x_B = AB \cos \phi \quad \text{and} \quad y_B = AB \sin \phi.$$

The MATLAB commands for joints A, E, and B are:

```
% position of joint A
xA = 0; yA = 0;
rA_  = [xA yA 0];
fprintf('Results \n\n')
fprintf('phi = phi1 = %g (degrees) \n',phi*180/pi)
fprintf('rA_ = [%g, %g, %g] (m)\n', rA_)
% position of joint E
```

```
xE =   a;
yE = -b;
rE_  = [xE yE 0]; % position vector of C
fprintf('rE_ = [%6.3f, %6.3f, %g] (m)\n', rE_)
% position of joint B
xB = AB*cos(phi);
yB = AB*sin(phi);
rB_  = [xB yB 0]; % position vector of B
fprintf('rB_ = [%6.3f, %6.3f, %g] (m)\n', rB_)
```

The numerical results are:

```
% rA_ = [0, 0, 0] (m)
% rE_ = [ 0.300, -0.250, 0] (m)
% rB_ = [ 0.141,  0.141, 0] (m)
```

Position of joint D
The unknowns are the x−coordinate of the joint D, x_D. The y−coordinate of D is
$y_D = c$. The coordinate x_D is calculated from the quadratic equation

$$(x_B - x_D)^2 + (y_B - y_D)^2 = BD^2,$$

or in MATLAB:

```
% position of joint D
yD = c;
syms xDs
eqD = (xB - xDs)^2 + (yB - yD)^2 - (BC+CD)^2;
solD = solve(eqD, 'xDs');
xD1=eval(solD(1));
xD2=eval(solD(2));
```

There are two solutions for x_D. For $\phi = 45°$ it is selected the solution $x_D > x_B$ and
in MATLAB:

```
if xD1 > xB xD = xD1; else xD = xD2; end
rD_  = [xD yD 0]; % position vector of D
fprintf('rD_ = [%6.3f, %g, %g] (m)\n', rD_)
```

The numerical value for the position vector of D is

```
% rD_ = [ 0.734, 0.05, 0] (m)
```

The angle of link 2 with the horizontal axis is calculated from the slope of the straight
line BD

$$\phi_2 = \arctan \frac{y_B - y_D}{x_B - x_D},$$

and in MATLAB:

```
% angle of link 2
phi2 = atan((yB-yD)/(xB-xD));
fprintf...
('phi2 = %6.3f (degrees) \n', phi2*180/pi)
% phi2 = -8.764 (degrees)
```

The coordinates of the joint C are calculated using two equations

$$(x_B - x_C)^2 + (y_B - y_C)^2 = BC^2 \quad \text{and} \quad \frac{y_B - y_D}{x_B - x_D} = \frac{y_D - y_C}{x_D - x_C},$$

or in MATLAB:

```
% position of joint C
xC=sym('xC','real');
yC=sym('yC','real');
eqC1 = (xB-xC)^2 + (yB-yC)^2 - BC^2;
eqC2 = (yB-yD)/(xB-xD)-(yD-yC)/(xD-xC);
solC = solve(eqC1, eqC2);
xCpos = eval(solC.xC);
yCpos = eval(solC.yC);
xC1 = xCpos(1); xC2 = xCpos(2);
yC1 = yCpos(1); yC2 = yCpos(2);
```

For $\phi = 45°$ the solution is selected using the condition $x_C > x_B$ and in MATLAB:

```
if xC1 > xB xC=xC1;yC=yC1; else
xC=xC2; yC=yC2; end
rC_ = [xC yC 0]; % position vector of C
fprintf('rC_ = [%6.3f, %6.3f, %g] (m)\n', rC_)
% rC_ = [ 0.349,  0.109, 0] (m)
```

The angle of link 4 with the horizontal axis is calculated from the slope of the line CE

$$\phi_4 = \arctan \frac{y_C - y_E}{x_C - x_E},$$

and in MATLAB:

```
% angle of link 4
phi4 = atan2((yC-yE),(xC-xE));
```

```
phi5 = phi4;
fprintf...
('phi4 = phi5 = %6.3f (degrees) \n', phi4*180/pi)
% phi4 = phi5 = 82.242 (degrees)
```

The coordinates of the point F are calculated with

$$x_F = x_C - CF \cos \phi_4 \quad \text{and} \quad y_F = y_C - CF \sin \phi_4.$$

The MATLAB commands for point F are:

```
xF = xC-CF*cos(phi4);
yF = yC-CF*sin(phi4);
rF_ = [xF yF 0];
fprintf('rF_ = [%6.3f, %6.3f, %g] (m)\n', rF_)
% rF_ = [ 0.288, -0.336, 0] (m)
```

The position of the mechanism for $\phi = 45°$ are obtained in MATLAB with:

```
% graphic of the mechanism
axis manual
hold on
grid on
sx = .8;
sy = .8;
axis([-sx sx -sy sy])
xlabel('x (m)'),ylabel('y (m)')
pM = plot([xA,xB],[yA,yB],'r-o',...
     [xB,xD],[yB,yD],'b-o',...
     [xC,xE],[yC,yE],'k-o',...
     [xC,xF],[yC,yF],'k-o');
plot(xC, yC, 'k.','Color', 'red')
xlabel('x (m)'),ylabel('y (m)')
pin0_scale = 0.015;
slider_scale = 0.05;
% plotting joint A
plot_pin0( xA, yA, pin0_scale, 0)
% plotting slider at D
plot_slider( xD, yD, slider_scale, pi)
% plotting slider at E
plot_slider( xE, yE, slider_scale, phi4)
% plotting joint E
plot_pivot0( xE, yE, pin0_scale, pi/2)
```

The mechanism plotted with MATLAB is shown in Fig. 4.2.

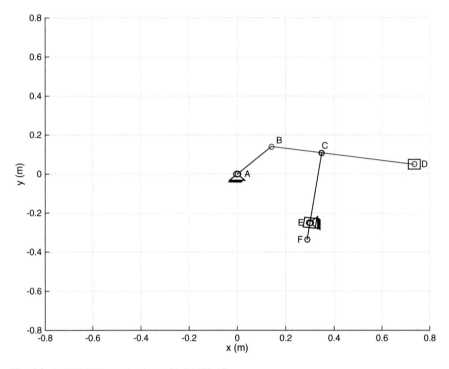

Fig. 4.2 R-RRT-RTR mechanism with MATLAB

4.2 Velocity and Acceleration Analysis

The velocity of the point B_1 on the link 1 is

$$\mathbf{v}_{B_1} = \mathbf{v}_A + \boldsymbol{\omega}_1 \times \mathbf{r}_{AB} = \boldsymbol{\omega}_1 \times \mathbf{r}_B,$$

where $\mathbf{v}_A \equiv \mathbf{0}$ is the velocity of the origin $A \equiv O$.

The angular velocity of link 1 is

$$\boldsymbol{\omega}_1 = \omega_1 \, \mathbf{k} = \frac{\pi n}{30} \, \mathbf{k} = \frac{\pi (-500)}{30} \, \mathbf{k} \ \text{rad/s}.$$

The velocity of point B_2 on the link 2 is $\mathbf{v}_{B_2} = \mathbf{v}_{B_1}$ because between the links 1 and 2 there is a rotational joint. The velocity of $B_1 = B_2$ is

$$\mathbf{v}_B = \mathbf{v}_{B_1} = \mathbf{v}_{B_2} = \begin{vmatrix} \mathbf{i} & \mathbf{j} & \mathbf{k} \\ 0 & 0 & \omega_1 \\ x_B & y_B & 0 \end{vmatrix}.$$

The acceleration of the point $B_1 = B_2$ is

$$\mathbf{a}_B = \mathbf{a}_{B_1} = \mathbf{a}_{B_2} = \mathbf{a}_A + \boldsymbol{\alpha}_1 \times \mathbf{r}_B + \boldsymbol{\omega}_1 \times (\boldsymbol{\omega}_1 \times \mathbf{r}_B) = \boldsymbol{\alpha}_1 \times \mathbf{r}_B - \omega_1^2 \mathbf{r}_B.$$

The angular acceleration of link 1 is $\boldsymbol{\alpha}_1 = \mathrm{d}\,\boldsymbol{\omega}_1 / \mathrm{d}\,t = \dot{\boldsymbol{\omega}}_1 = \mathbf{0}$. The MATLAB commands for the velocity and acceleration of B are:

```
n = -500; % rpm
omega1_ = [0 0 pi*n/30]; alpha1_ = [0 0 0];
fprintf...
('omega1_ = [%g, %g, %6.3f] (rad/s)\n', omega1_)
fprintf...
('alpha1_ = [%g, %g, %g] (rad/s^2)\n', alpha1_)
fprintf('\n')

vA_ = [0 0 0 ]; aA_ = [0 0 0 ];

vB1_ = vA_ + cross(omega1_,rB_);
vB2_ = vB1_;
aB1_ = aA_ + cross(alpha1_,rB_) - ...
    dot(omega1_,omega1_)*rB_;
vB2_ = vB1_; vB_ = vB1_;
aB2_ = aB1_; aB_ = aB1_;
fprintf...
('vB_ = vB1_ = vB2_ = [%6.3f, %6.3f, %g] (m/s)\n',vB1_)
fprintf...
('aB_ = aB1_ = aB2_ = [%6.3f, %6.3f, %g] (m/s^2)\n',aB1_)
%
% omega1_ = [0, 0, -52.360] (rad/s)
% alpha1_ = [0, 0, 0] (rad/s^2)
%
% vB_ = vB1_ = vB2_ = [ 7.405, -7.405, 0] (m/s)
% aB_ = aB1_ = aB2_ = [-387.715, -387.715, 0] (m/s^2)
```

The velocity of the point D_2 on the link 2 is calculated in terms of the velocity of the point B_2 on the link 2

$$\mathbf{v}_{D_2} = \mathbf{v}_{B_2} + \boldsymbol{\omega}_2 \times \mathbf{r}_{BD} = \mathbf{v}_B + \boldsymbol{\omega}_2 \times (\mathbf{r}_D - \mathbf{r}_B). \qquad (4.1)$$

where $\boldsymbol{\omega}_2 = \omega_2\,\mathbf{k}$ is the unknown angular velocity of link 2. This velocity of D on link 2 is equal with the velocity of D on slider 3. The slider has a translation motion along the $x-$axis and

$$\mathbf{v}_{D_2} = \mathbf{v}_{D_3} = v_D\,\mathbf{l}, \qquad (4.2)$$

where v_D is unknown. Equations (4.1) and (4.2) represent a system of two equations with two unknowns ω_2 and v_D

$$v_D \mathbf{1} = \mathbf{v}_B + \begin{vmatrix} \mathbf{1} & \mathbf{J} & \mathbf{k} \\ 0 & 0 & \omega_2 \\ x_D - x_B & y_D - y_B & 0 \end{vmatrix},$$

or

$$v_D = v_{Bx} - \omega_2 (y_D - y_B),$$
$$0 = v_{By} + \omega_2 (x_D - x_B). \tag{4.3}$$

The MATLAB commands for calculating ω_2 and v_D are

```
omega2z=sym('omega2z','real');
vD=sym('vD','real');
omega2_ = [ 0 0 omega2z ];
% vD2_ = vB_ + omegaB_ x rBD_
vD2_ = vB1_ + cross(omega2_,rD_-rB_);
% vD3 is parallel to the sliding direction x
vD3_ = [vD 0 0];
eqvD_ = vD3_ - vD2_;  % vectorial equation
% the component of the vectorial equation on x-axis
eqvDx = eqvD_(1);
% the component of the vectorial equation on y-axis
eqvDy = eqvD_(2);
% two equations eqvBx and eqvBy with two unknowns
% solve for omega2z and vD
solvD = solve(eqvDx,eqvDy);
omega2=eval(solvD.omega2z);
vDs=eval(solvD.vD);
omega2_ = [0 0 omega2];
vD_ = [vDs 0 0];
% print the equations for calculating  omega2 and vD
fprintf...
('vD2_ = vB_ + omega2_ x rBD_ = vD3_   => \n')
digits(3)
qvDx=vpa(eqvDx);
fprintf('x-axis: %s = 0 \n', char(qvDx))
qvDy=vpa(eqvDy);
fprintf('y-axis: %s = 0 \n', char(qvDy))
fprintf('=>\n')
fprintf('omega2z = %6.3f (rad/s)\n', omega2)
fprintf('vD = %6.3f (m/s)\n', vDs)
fprintf('\n')
fprintf...
('omega2_ = [%g,%g,%6.3f](rad/s)\n', omega2_)
fprintf('vD_ = [%6.3f,%6.3f,%d] (m/s)\n\n', vD_)
```

```
%
% omega2_ = [0,0,12.487](rad/s)
% vD_ = [ 8.546, 0.000,0]  (m/s)
```

The acceleration of the point D_2 on the link 2 is calculated in terms of the acceleration of the point B_2 on the link 2

$$\mathbf{a}_{D_2} = \mathbf{a}_{B_2} + \boldsymbol{\alpha}_2 \times \mathbf{r}_{BD} + \boldsymbol{\omega}_2 \times (\boldsymbol{\omega}_2 \times \mathbf{r}_{BD}).$$ (4.4)

where $\boldsymbol{\alpha}_2 = \alpha_2 \mathbf{k}$ is the unknown angular acceleration of link 2. This acceleration of D on link 2 is equal with the acceleration of D on slider 3. The slider has a translation motion along the $x-$axis and

$$\mathbf{a}_{D_2} = \mathbf{a}_{D_3} = a_D \mathbf{1},$$ (4.5)

where a_D is unknown. Equations (4.4) and (4.5) represent a system of two equations with two unknowns α_2 and a_D

$$a_D \mathbf{1} = \mathbf{a}_B + \begin{vmatrix} \mathbf{1} & \mathbf{J} & \mathbf{k} \\ 0 & 0 & \alpha_2 \\ x_D - x_B & y_D - y_B & 0 \end{vmatrix} - \omega_2^2 \mathbf{r}_{BD},$$

or

$$a_D = a_{Bx} - \alpha_2 (y_D - y_B) - \omega_2^2 (x_D - x_B),$$
$$0 = a_{By} + \alpha_2 (x_D - x_B) - \omega_2^2 (y_D - y_B).$$ (4.6)

The angular acceleration ω_2 and the acceleration a_D are calculated with the following MATLAB commands:

```
alpha2z=sym('alpha2z','real');
aD=sym('aD','real');
alpha2_ = [ 0 0 alpha2z ];
aD2_ = aB_ + cross(alpha2_, rD_-rB_) -...
    dot(omega2_, omega2_)*(rD_-rB_);
aD3_ = [aD 0 0];
eqaD_ = aD3_ - aD2_; % vectorial equation
eqaDx = eqaD_(1); % equation component on x-axis
eqaDy = eqaD_(2); % equation component on y-axis
solaD = solve(eqaDx,eqaDy);
alpha2zs=eval(solaD.alpha2z);
aDs=eval(solaD.aD);
alpha2_ = [0 0 alpha2zs];
aD_ = [aDs 0 0];
```

```
% alpha2_ = [0,0,629.786](rad/s^2)
% aD_ = [-422.604, 0.000,0] (m/s^2)
```

The velocity and acceleration of the joint C are calculated with

$$\mathbf{v}_C = \mathbf{v}_B + \boldsymbol{\omega}_2 \times \mathbf{r}_{BC},$$

$$\mathbf{a}_C = \mathbf{a}_B + \boldsymbol{\alpha}_2 \times \mathbf{r}_{BC} + \boldsymbol{\omega}_2 \times (\boldsymbol{\omega}_2 \times \mathbf{r}_{BC}),$$

or with MATLAB:

```
vC_ = vB_ + cross(omega2_,rC_-rB_);
aC_ = aB_ + cross(alpha2_, rC_-rB_) + ...
    cross(omega2_,cross(omega2_, rC_-rB_));

% vC_=vC2_=vC4_=[ 7.804,-4.813,0] (m/s)
% aC_=aC2_=aC4_=[-399.926,-252.015,0] (m/s^2)
```

The velocity of the point E_4 on the link 4 is calculated in terms of the velocity of the point C

$$\mathbf{v}_{E_4} = \mathbf{v}_C + \boldsymbol{\omega}_4 \times \mathbf{r}_{CE}, \tag{4.7}$$

where the unknown angular velocity of link 4 is $\boldsymbol{\omega}_4 = \omega_4\,\mathbf{k}$.

The velocity of the point E_4 on the link 4 is also calculated in terms of the velocity of the point E_5 on the link 5

$$\mathbf{v}_{E_4} = \mathbf{v}_{E_5} + \mathbf{v}_{E_{45}}^{rel} = \mathbf{0} + \mathbf{v}_{E_{45}}, \tag{4.8}$$

where $\mathbf{v}_{E_5} = \mathbf{0}$ and $\mathbf{v}_{E_{45}}^{rel} = \mathbf{v}_{E_{45}}$ is the relative velocity of E_4 with respect to a reference frame attached to link 5. This relative velocity is parallel to the sliding direction CF, $\mathbf{v}_{E_{45}} || CF$, or

$$\mathbf{v}_{E_{45}} = v_{E_{45}} \cos\phi_4\,\mathbf{I} + v_{E_{45}} \sin\phi_4\,\mathbf{J}, \tag{4.9}$$

where the angle ϕ_4 is known from the position analysis. Equations (4.7), (4.8), and (4.9) give

$$\mathbf{v}_C + \begin{vmatrix} \mathbf{I} & \mathbf{J} & \mathbf{k} \\ 0 & 0 & \omega_4 \\ x_E - x_C & y_E - y_C & 0 \end{vmatrix} = v_{E_{45}} \cos\phi_4\,\mathbf{I} + v_{E_{45}} \sin\phi_4\,\mathbf{J}. \tag{4.10}$$

Equation (4.10) represents a vectorial equation with two scalar components on x-axis and y-axis and with two unknowns ω_4 and $v_{E_{45}}$

$$v_{C_x} - \omega_4 (y_E - y_C) = v_{E_{45}} \cos \phi_4,$$
$$v_{C_y} + \omega_4 (x_E - x_C) = v_{E_{45}} \sin \phi_4.$$

The MATLAB program for ω_4 and $v_{E_{45}}$ is

```
omega4z=sym('omega4z','real');
vE45=sym('vE45','real');
omega4_ = [ 0 0 omega4z ];
vE5_ = [0 0 0 ]; % E is fixed
vE4_ = vC_ + cross(omega4_,rE_-rC_);
vE4E5_ = vE45*[ cos(phi4) sin(phi4) 0];
eqvE_ = vE4_ - vE5_ - vE4E5_; % vectorial equation
eqvEx = eqvE_(1);
eqvEy = eqvE_(2);
solvE = solve(eqvEx,eqvEy);
omega4zs=eval(solvE.omega4z);
vE45s=eval(solvE.vE45);

omega4_ = [0 0 omega4zs];
omega5_ = omega4_;
vE45_ = vE45s*[cos(phi4) sin(phi4) 0];

% omega4_=omega5_ = [0,0,-23.109](rad/s)
% vE45_ = [-0.502,-3.681,0] (m/s)
```

The acceleration of the point E_4 on the link 4 is calculated in terms of the acceleration of the point C

$$\mathbf{a}_{E_4} = \mathbf{a}_C + \boldsymbol{\alpha}_4 \times \mathbf{r}_{CE} + \boldsymbol{\omega}_4 \times (\boldsymbol{\omega}_4 \times \mathbf{r}_{CE}), \quad (4.11)$$

where the unknown angular acceleration of link 4 is $\boldsymbol{\alpha}_4 = \alpha_4 \mathbf{k}$.

The acceleration of the point E_4 on the link 4 is also calculated in terms of the acceleration of the point E_5 on the link 5

$$\mathbf{a}_{E_4} = \mathbf{a}_{E_5} + \mathbf{a}_{E_{45}}^{rel} + 2\,\boldsymbol{\omega}_4 \times \mathbf{v}_{E_{45}} = \mathbf{0} + \mathbf{a}_{E_{45}} + 2\,\boldsymbol{\omega}_4 \times \mathbf{v}_{E_{45}}, \quad (4.12)$$

where $\mathbf{a}_{E_5} = \mathbf{0}$ and $\mathbf{a}_{E_{45}}^{rel} = \mathbf{a}_{E_{45}}$ is the relative acceleration of E_4 with respect to a reference frame attached to link 5. This relative acceleration is parallel to the sliding direction CF, $\mathbf{a}_{E_{45}}||CF$, or

$$\mathbf{a}_{E_{45}} = a_{E_{45}} \cos \phi_4 \mathbf{I} + a_{E_{45}} \sin \phi_4 \mathbf{J}. \quad (4.13)$$

The Coriolis acceleration of E_4 relative to E_5 is

$$\mathbf{a}_{E_{45}}^{cor} = 2\,\boldsymbol{\omega}_4 \times \mathbf{v}_{E_{45}} = 2 \begin{vmatrix} \mathbf{1} & \mathbf{J} & \mathbf{k} \\ 0 & 0 & \omega_4 \\ v_{E_{45}}\cos\phi_4 & v_{E_{45}}\sin\phi_4 & 0 \end{vmatrix}. \qquad (4.14)$$

Equations (4.11), (4.12), (4.13), and (4.14) give

$$\mathbf{a}_C + \boldsymbol{\alpha}_4 \times \mathbf{r}_{CE} - \omega_4^2\,\mathbf{r}_{CE} = a_{E_{45}}\cos\phi_4\,\mathbf{1} + a_{E_{45}}\sin\phi_4\,\mathbf{J} + 2\,\boldsymbol{\omega}_4 \times \mathbf{v}_{E_{45}}. \quad (4.15)$$

Equation (4.15) represents a vectorial equation with two scalar components on x-axis and y-axis and with two unknowns α_4 and $a_{E_{45}}$. The MATLAB commands for calculating α_4 and $a_{E_{45}}$ are

```
aE45cor_ = 2*cross(omega4_,vE45_); % Coriolis acceleration
alpha4z=sym('alpha4z','real');
aE45=sym('aE45','real');
alpha4_ = [ 0 0 alpha4z ];
aE5_ = [0 0 0 ]; % E is fixed
aE4_=aC_+cross(alpha4_,rE_-rC_)+...
cross(omega4_,cross(omega4_,rE_-rC_));
aE4E5_ = aE45*[ cos(phi4) sin(phi4) 0];
eqaE_ = aE4_ - aE5_ - aE4E5_ - aE45cor_; % vectorial equation
eqaEx = eqaE_(1);
eqaEy = eqaE_(2);
% x-axis: 0.359*alpha4z - 0.135*aE45 - 204.0 = 0
% y-axis: - 0.991*aE45 - 0.049*alpha4z - 83.3 = 0
solaE = solve(eqaEx,eqaEy);
alpha4zs=eval(solaE.alpha4z);
aE45s=eval(solaE.aE45);
alpha4_ = [0 0 alpha4zs]; alpha5_=alpha4_;
aE45_ = aE45s*[ cos(phi4) sin(phi4) 0];

% alpha4_=alpha5_ = [0,0,525.219](rad/s^2)
% aE45_ = [-14.847,-108.974,0] (m/s^2)
```

The velocity and acceleration of the end point F are calculated with

$$\mathbf{v}_F = \mathbf{v}_C + \boldsymbol{\omega}_4 \times \mathbf{r}_{CF},$$
$$\mathbf{a}_F = \mathbf{a}_C + \boldsymbol{\alpha}_4 \times \mathbf{r}_{CF} + \boldsymbol{\omega}_4 \times (\boldsymbol{\omega}_4 \times \mathbf{r}_{CF}),$$

or with MATLAB:

```
vF_ = vC_ + cross(omega4_,rF_-rC_);
aF_=aC_+cross(alpha4_,rF_-rC_)+...
cross(omega4_,cross(omega4_,rF_-rC_));
%
```

```
% vF_=[-2.500,-3.409,0]  (m/s)
% aF_=[-133.299,-45.808,0]  (m/s^2)
```

4.3   Dynamic Force Analysis

For the force analysis the positions and the accelerations of the mass centers of the
links are calculated.
    The position vector of the center of mass of link 1 is

$$\mathbf{r}_{C_1} = x_{C_1}\,\mathbf{1} + y_{C_1}\,\mathbf{J} = \frac{x_B}{2}\,\mathbf{1} + \frac{y_B}{2}\,\mathbf{J}.$$

The position vector of the center of mass of link 2 is

$$\mathbf{r}_{C_2} = x_{C_2}\,\mathbf{1} + y_{C_2}\,\mathbf{J} = \frac{x_B + x_D}{2}\,\mathbf{1} + \frac{y_B + y_D}{2}\,\mathbf{J}.$$

The position vector of the center of mass of slider 3 is

$$\mathbf{r}_{C_3} = x_{C_3}\,\mathbf{1} + y_{C_3}\,\mathbf{J} = x_D\,\mathbf{1} + y_D\,\mathbf{J}.$$

The position vector of the center of mass of link 4 is

$$\mathbf{r}_{C_4} = x_{C_4}\,\mathbf{1} + y_{C_4}\,\mathbf{J} = \frac{x_C + x_F}{2}\,\mathbf{1} + \frac{y_C + y_F}{2}\,\mathbf{J}.$$

The position vector of the center of mass of slider 5 is

$$\mathbf{r}_{C_5} = x_{C_5}\,\mathbf{1} + y_{C_5}\,\mathbf{J} = \mathbf{r}_E.$$

The MATLAB commands for the position vectors of the mass centers are:

```
rC1_ = rB_/2;
rC2_ = (rB_+rD_)/2;
rC3_ = rD_;
rC4_ = (rC_+rF_)/2;
rC5_ = rE_;
% rC1_ = [ 0.071,  0.071,0]  (m)
% rC2_ = [ 0.438,  0.096,0]  (m)
% rC3_ = [ 0.734,  0.050,0]  (m)
% rC4_ = [ 0.319,-0.114,0]  (m)
% rC5_ = [ 0.300,-0.250,0]  (m)
```

The acceleration vector of the center of mass of link 1 is

$$\mathbf{a}_{C_1} = \frac{\mathbf{a}_B}{2}.$$

The acceleration vector of the center of mass of link 2 is

$$\mathbf{a}_{C_2} = \frac{\mathbf{a}_B + \mathbf{a}_D}{2}.$$

The acceleration vector of the center of mass of slider 3 is

$$\mathbf{a}_{C_3} = \mathbf{a}_D.$$

The acceleration vector of the center of mass of link 4 is

$$\mathbf{a}_{C_4} = \frac{\mathbf{a}_C + \mathbf{a}_F}{2}.$$

The acceleration vector of the center of mass of slider 5 is

$$\mathbf{a}_{C_5} = \mathbf{a}_E.$$

The MATLAB commands for the acceleration of the mass centers are:

```
aC1_ = aB_/2;
aC2_ = (aB_+aD_)/2;
aC3_ = aD_;
aC4_ = (aC_+aF_)/2;
aC5_ = [0 0 0];

% aC1_ = [-193.857,-193.857,0] (m/s^2)
% aC2_ = [-405.159,-193.857,0] (m/s^2)
% aC3_ = [-422.604, 0.000,0] (m/s^2)
% aC4_ = [-266.612,-148.911,0] (m/s^2)
% aC5_ = [ 0.000,0,0] (m/s^2)
```

The angular accelerations of the links 1, 2, 3, 4, and 5 are

```
alpha1_ = [0,0, 0.000](rad/s^2)
alpha2_ = [0,0,629.786](rad/s^2)
alpha3_ = [0,0, 0.000](rad/s^2)
alpha4_ = [0,0,525.219](rad/s^2)
alpha5_ = [0,0,525.219](rad/s^2)
```

The height and the depth of the straight links 1, 2, and 4 are:

```
h = 0.02; % (m) height of the links
d = 0.01; % (m) depth of the links
```

The height, width, and depth of the sliders 3 and 5 are:

```
hSlider = 0.04; % (m) height of the slider
wSlider = 0.08; % (m) width of the slider
dSlider = 0.02; % (m) depth of the slider
```

The links are homogeneous with the mass density rho and the gravitational acceleration is g:

```
rho = 8000; % (kg/m^3) density of the material
g = 9.807; % (m/s^2) gravitational acceleration
```

An external moment acts on slider 5, $\mathbf{M}_{5\text{ext}} = \mathbf{M}_e$, and is opposed to the motion of the slider:

```
Me=1000; % N m
Me_ = -sign(omega5_(3))*[0,0,Me];
% Me_ = [0, 0, 1000.000] (N m)
```

The mass of the crank 1 is $m_1 = \rho\, AB\, h\, d$, the inertia force on link 1 at C_1 is $\mathbf{F}_{\text{in}\,1} = -m_1\,\mathbf{a}_{C_1}$, the gravitational force on crank 1 at C_1 is $\mathbf{G}_1 = -m_1\, g\,\mathbf{J}$, the mass moment of inertia of the link 1 about C_1 is $I_{C_1} = m_1\, AB^2/12$, the moment of inertia on link 1 is $\mathbf{M}_{\text{in}\,1} = -I_{C_1}\,\boldsymbol{\alpha}_1 = \mathbf{0}$, and with MATLAB:

```
fprintf('Link 1   \n')
m1 = rho*AB*h*d;
Fin1_ = -m1*aC1_;
G1_ = [0,-m1*g,0];
IC1 = m1*AB^2/12;
alpha1_ = [0 0 0];
Min1_ = -IC1*alpha1_;
% m1 = 0.320 (kg)
% m1 aC1_ = [-62.034,-62.034,0] (N)
% Fin1_= - m1 aC1_ = [62.034,62.034,0] (N)
% G1_ = - m1 g = [0,-3.138,0] (N)
% IC1 = 0.00106667 (kg m^2)
% IC1 alpha1_=[0, 0, 0](N m)
% Min1_=-IC1 alpha1_=[0, 0, 0](N m)
```

The mass of link 2 is $m_2 = \rho\,(BC + CD)\,h\,d$, the inertia force on link 2 at C_2 is $\mathbf{F}_{\text{in}\,2} = -m_2\,\mathbf{a}_{C_2}$, the gravitational force at C_2 is $\mathbf{G}_2 = -m_2\, g\,\mathbf{J}$, the mass moment of inertia of the link about C_2 is $I_{C_2} = m_1\,(BC + CD)^2/12$, the moment of inertia is $\mathbf{M}_{\text{in}\,2} = -I_{C_2}\,\boldsymbol{\alpha}_2$, and with MATLAB:

```
fprintf('Link 2   \n')
```

```
m2 = rho*(BC+CD)*h*d;
Fin2_ = -m2*aC2_;
G2_  = [0,-m2*g,0];
IC2 = m2*(BC+CD)^2/12;
Min2_ = -IC2*alpha2_;
% m2 =  0.960 (kg)
% m2 aC2_ = [-388.953,-186.103,0] (N)
% Fin2_= - m2 aC2_ = [388.953,186.103,0] (N)
% G2_ = - m2 g = [0,-9.415,0] (N)
% IC2 = 0.0288 (kg m^2)
% IC2 alpha2_=[ 0, 0,1.813784e+01](N m)
% Min2_=-IC2 alpha2_=[0, 0, -1.813784e+01](N m)
```

The mass of slider 3 is $m_3 = \rho\, h_{\text{slider}}\, w_{\text{slider}}\, d_{\text{slider}}$, the inertia force on link 3 at $C_3 = D$ is $\mathbf{F}_{\text{in}\,3} = -m_3\, \mathbf{a}_{C_3}$, the gravitational force at C_3 is $\mathbf{G}_3 = -m_3\, g\, \mathbf{J}$, the mass moment of inertia of the link about C_3 is $I_{C_3} = m_3\, (h^2_{\text{slider}} + w^2_{\text{slider}})/12$, the moment of inertia is $\mathbf{M}_{\text{in}\,3} = -I_{C_3}\, \boldsymbol{\alpha}_3 = \mathbf{0}$, and with MATLAB:

```
m3 = rho*hSlider*wSlider*dSlider;
Fin3_ = -m3*aC3_;
G3_  = [0,-m3*g,0];
IC3 = m3*(hSlider^2+wSlider^2)/12;
Min3_ = -IC3*alpha3_;
% m3 =  0.512 (kg)
% m3 aC3_ = [-216.373, 0, 0] (N)
% Fin3_= - m3 aC3_ =[216.373, 0, 0] (N)
% G3_ = - m3 g_ = [0,-5.021,0] (N)
% IC3 = 3.413e-04 (kg m^2)
% IC3 alpha3_=[0, 0, 0](N m)
% Min3_=-IC3 alpha3_=[0, 0, 0](N m)
```

The mass of link 4 is $m_4 = \rho\, C F\, h\, d$, the inertia force on link 4 at C_4 is $\mathbf{F}_{\text{in}\,4} = -m_4\, \mathbf{a}_{C_4}$, the gravitational force at C_4 is $\mathbf{G}_4 = -m_4\, g\, \mathbf{J}$, the mass moment of inertia of the link about C_4 is $I_{C_4} = m_4\, C F^2/12$, the moment of inertia is $\mathbf{M}_{\text{in}\,4} = -I_{C_4}\, \boldsymbol{\alpha}_4$, and with MATLAB:

```
m4 = rho*CF*h*d;
Fin4_ = -m4*aC4_;
G4_  = [0,-m4*g,0];
IC4 = m4*CF^2/12;
Min4_ = -IC4*alpha4_;
% m4 =  0.720 (kg)
% m4 aC4_ = [-1066.450,-595.645,0] (N)
% Fin4_= - m4 aC4_ =[191.961,107.216,0] (N)
% G4_ = - m4 g_ = [0,-7.061,0] (N)
```

```
% IC4 = 1.215e-02 (kg m^2)
% IC4 alpha4_=[0,0,6.381](N m)
% Min4_=-IC4 alpha4_=[0,0,-6.381](N m)
```

The mass of slider 5 is $m_5 = \rho\, h_{\text{slider}}\, w_{\text{slider}}\, d_{\text{slider}}$, the inertia force on link 5 at $C_5 = E$ is $\mathbf{F}_{\text{in}\,5} = -m_5\, \mathbf{a}_{C_5}$, the gravitational force at C_5 is $\mathbf{G}_5 = -m_5\, g\, \mathbf{J}$, the mass moment of inertia of the link about C_5 is $I_{C_5} = m_5\,(h^2_{\text{slider}} + w^2_{\text{slider}})/12$, the moment of inertia is $\mathbf{M}_{\text{in}\,5} = -I_{C_5}\, \boldsymbol{\alpha}_5$, and with MATLAB:

```
m5 = rho*hSlider*wSlider*dSlider;
Fin5_ = -m5*aC5_;
G5_ = [0,-m5*g,0];
IC5 = m5*(hSlider^2+wSlider^2)/12;
Min5_ = -IC5*alpha5_;
% m5 =   0.512 (kg)
% m5 aC5_ = [0,0,0] (N)
% Fin5_ = - m5 aC5 =[-0,-0,0] (N)
% G5_ = - m5 g = [0, -5.021, 0] (N)
% IC5 = 3.413e-04 (kg m^2)
% IC5 alpha5_=[0, 0,  0.179](N m)
% Min5_=-IC5 alpha5_=[0, 0, -0.179](N m)
```

The dynamic force analysis starts with the last link 5 because the given external moment acts on this link. The free-body diagram of the slider 5 is shown in Fig. 4.3a. The unknown joint reaction force of the ground on the slider 5 acts at E, $\mathbf{F}_{05} = F_{05x}\, \mathbf{1} + F_{05y}\, \mathbf{J}$, and in MATLAB is:

```
F05x=sym('F05x','real');
F05y=sym('F05y','real');
F05_=[F05x, F05y, 0]; % unknown joint force
```

The unknown joint reaction force of the link 4 on the slider 5 acts at E, $\mathbf{F}_{45} = F_{45x}\, \mathbf{1} + F_{45y}\, \mathbf{J}$, and in MATLAB is:

```
F45x=sym('F45x','real');
F45y=sym('F45y','real');
F45_=[F45x, F45y, 0]; % unknown joint force
```

The unknown joint reaction moment of the link 4 on the slider 5 is $\mathbf{M}_{45} = M_{45z}\, \mathbf{k}$, and in MATLAB is:

```
M45z=sym('M45z','real');
M45_=[0, 0, M45z]; % unknown joint moment
```

Fig. 4.3 Free-body
diagrams for links 4 and 5

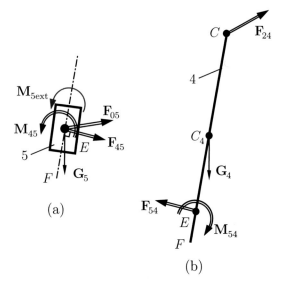

(a)

(b)

Because the joint between 4 and 5 at E is a translational joint the reaction force $\mathbf{F}_{45}$, (F45_), is perpendicular to the sliding direction CE

$$\mathbf{F}_{45} \cdot \mathbf{r}_{CE} = 0 \quad \text{or} \quad (F_{45x}\,\mathbf{1} + F_{45y}\,\mathbf{J}) \cdot [(x_E - x_C)\,\mathbf{1} + (y_E - y_C)\,\mathbf{J}] = 0, \quad (4.16)$$

and with MATLAB:

```
eqF45CE = dot(F45_,rE_-rC_); % (1)
```

The Newton vectorial equation of motion for slider 5 is

$$m_5\,\mathbf{a}_{C_5} = \sum \mathbf{F}^{(5)} = \mathbf{F}_{45} + \mathbf{F}_{05} + \mathbf{G}_5, \tag{4.17}$$

or

$$m_5\,a_{C_5x} = F_{45x} + F_{05x}, \tag{4.18}$$

$$m_5\,a_{C_5y} = F_{45y} + F_{05y} - m_5\,g, \tag{4.19}$$

and with MATLAB:

```
eqF5_  = -m5*aC5_+F45_+F05_+G5_;
eqF5x = eqF5_(1); % (2) x-component
eqF5y = eqF5_(2); % (3) y-component
```

The Euler vectorial equation of motion for slider 5 gives

$$I_{C_5}\,\boldsymbol{\alpha}_5 = \sum \mathbf{M}_{C_5}^{(5)} = \mathbf{M}_{45} + \mathbf{M}_e, \qquad (4.20)$$

or

$$I_{C_5}\,\alpha_5\,\mathbf{k} = M_{45z}\,\mathbf{k} + M_{ez}\,\mathbf{k}, \qquad (4.21)$$

and with MATLAB:

```
% sum M of 5 wrt E
eqMC5_  = -IC5*alpha5_+M45_+Me_;
eqMC5z = eqMC5_(3); % (4) z-component
```

There are four scalar equations Eqs. (4.16), (4.18), (4.19), (4.21) with five unknowns F_{05x}, F_{05y}, F_{45x}, F_{45y}, M_{45z}.

The force calculation will continue with link 4 as shown in Fig. 4.3b. The reaction of link 2 on link 4 at C is $\mathbf{F}_{24} = F_{24x}\,\mathbf{\imath} + F_{24y}\,\mathbf{\jmath}$ and with MATLAB:

```
F24x=sym('F24x','real');
F24y=sym('F24y','real');
F24_=[F24x, F24y, 0]; % unknown joint force
```

For link 4 the Newton-Euler equations of motion are

$$m_4\,\mathbf{a}_{C_4} = \sum \mathbf{F}^{(4)} = \mathbf{F}_{54} + \mathbf{F}_{24} + \mathbf{G}_4, \qquad (4.22)$$

$$I_{C_4}\,\boldsymbol{\alpha}_4 = \sum \mathbf{M}_{C_4}^{(4)} = \mathbf{M}_{54} + \mathbf{r}_{C_4C} \times \mathbf{F}_{24} + \mathbf{r}_{C_4E} \times \mathbf{F}_{54}. \qquad (4.23)$$

The scalar equations of motion for the link 4 are

$$m_4\,a_{C_4x} = \mathbf{F}_{54x} + \mathbf{F}_{24x}, \qquad (4.24)$$

$$m_4\,a_{C_4y} = \mathbf{F}_{54y} + \mathbf{F}_{24y} - m_4\,g, \qquad (4.25)$$

$$I_{C_4}\,\alpha_4\,\mathbf{k} = M_{54z}\,\mathbf{k}$$

$$+ \begin{vmatrix} \mathbf{\imath} & \mathbf{\jmath} & \mathbf{k} \\ x_C - x_{C_4} & y_C - y_{C_4} & 0 \\ F_{24x} & F_{24y} & 0 \end{vmatrix} + \begin{vmatrix} \mathbf{\imath} & \mathbf{\jmath} & \mathbf{k} \\ x_E - x_{C_4} & y_E - y_{C_4} & 0 \\ F_{54x} & F_{54y} & 0 \end{vmatrix}. \qquad (4.26)$$

The MATLAB commands for the equations of motion of the link 4 are:

```
F54_=-F45_;
M54_=-M45_;
% sum F of 5
eqF4_  = -m4*aC4_+F54_+F24_+G4_;
eqF4x = eqF4_(1); % (5) x-component
eqF4y = eqF4_(2); % (6) y-component
```

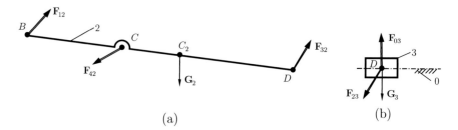

Fig. 4.4 Free-body diagrams for links 2 and 3

```
% sum M of 4 wrt C4
eqMC4_ = -IC4*alpha4_+cross(rC_-rC4_,F24_)...
          +cross(rE_-rC4_,F54_)+M54_;
eqMC4z = eqMC4_(3); % (7) z-component
```

For the links 5 and 4 there are seven equations Eqs. (4.16), (4.18), (4.19), (4.21), (4.24), (4.25), (4.26) with seven unknowns F_{05x}, F_{05y}, F_{45x}, F_{45y}, M_{45z}, F_{24x}, F_{24y}. The algebraic equations are solved using the MATLAB relations:

```
sol45=solve(eqF45CE,eqF5x,eqF5y,eqMC5z,eqF4x,eqF4y,eqMC4z);

F24xs=eval(sol45.F24x);
F24ys=eval(sol45.F24y);
F05xs=eval(sol45.F05x);
F05ys=eval(sol45.F05y);
F45xs=eval(sol45.F45x);
F45ys=eval(sol45.F45y);
M45zs=eval(sol45.M45z);

F24_ = [F24xs F24ys 0];
F05_ = [F05xs F05ys 0];
F45_ = [F45xs F45ys 0];
M45_ = [0 0 M45zs];
```

and the results are:

```
% F05_ = [-2822.194,389.528,0] (N)
% F24_ = [2630.233,-484.662,0] (N)
% F45_ = [2822.194,-384.507,0] (N)
% M45_ = [0,0,-999.821] (N m)
```

The dynamic force analysis will consider the link 2 as shown in Fig. 4.4a.

The reaction force of link 1 on link 2 at B is $\mathbf{F}_{12} = F_{12x}\,\mathbf{i} + F_{12y}\,\mathbf{j}$ and with MATLAB:

```
F12x=sym('F12x','real');
F12y=sym('F12y','real');
F12_=[F12x, F12y, 0]; % unknown joint force
```

The reaction force of slider 3 on link 2 at D is $\mathbf{F}_{32} = F_{32x}\,\mathbf{1} + F_{32y}\,\mathbf{J}$ and with MAT-LAB:

```
F32x=sym('F32x','real');
F32y=sym('F32y','real');
F32_=[F32x, F32y, 0]; % unknown joint force
```

For link 2 the Newton-Euler equations of motion are

$$m_2\,\mathbf{a}_{C_2} = \sum \mathbf{F}^{(2)} = \mathbf{F}_{42} + \mathbf{F}_{12} + \mathbf{F}_{32} + \mathbf{G}_2, \tag{4.27}$$

$$I_{C_2}\,\alpha_2 = \sum \mathbf{M}_{C_2}^{(2)} = \mathbf{r}_{C_2B} \times \mathbf{F}_{12} + \mathbf{r}_{C_2D} \times \mathbf{F}_{32} + \mathbf{r}_{C_2C} \times \mathbf{F}_{42}. \tag{4.28}$$

The scalar equations of motion for the link 4 are

$$m_2\,a_{C_2x} = F_{42x} + F_{12x} + F_{32x}, \tag{4.29}$$

$$m_2\,a_{C_2y} = F_{42y} + F_{12y} + F_{32y} - m_2\,g, \tag{4.30}$$

$$I_{C_2}\,\alpha_2\,\mathbf{k} = \begin{vmatrix} \mathbf{1} & \mathbf{J} & \mathbf{k} \\ x_B - x_{C_2} & y_B - y_{C_2} & 0 \\ F_{12x} & F_{12y} & 0 \end{vmatrix}$$

$$+ \begin{vmatrix} \mathbf{1} & \mathbf{J} & \mathbf{k} \\ x_D - x_{C_2} & y_D - y_{C_2} & 0 \\ F_{32x} & F_{32y} & 0 \end{vmatrix} + \begin{vmatrix} \mathbf{1} & \mathbf{J} & \mathbf{k} \\ x_C - x_{C_2} & y_C - y_{C_2} & 0 \\ F_{42x} & F_{42y} & 0 \end{vmatrix}. \tag{4.31}$$

The MATLAB commands for the equations of motion of the link 2 are:

```
% sum F of 2
F42_ = -F24_;
eqF2_ = -m2*aC2_+F42_+F12_+F32_+G2_;
eqF2x = eqF2_(1); % (8) x-component
eqF2y = eqF2_(2); % (9) y-component
% sum M of 2 wrt C2
eqMC2_ = -IC2*alpha2_+cross(rB_-rC2_,F12_)...
         +cross(rD_-rC2_,F32_)+cross(rC_-rC2_,F42_);
eqMC2z = eqMC2_(3); % (10)
```

For the link 2 there are three equations Eqs. (4.29), (4.30), (4.31) with four unknowns F_{12x}, F_{12y}, F_{32x}, F_{32y}. Next the slider 3 is considered as shown in Fig. 4.4b. The reaction force of the ground on link 3 at D is perpendicular to the sliding direction, $\mathbf{F}_{03} = F_{03y}\,\mathbf{J}$, and with MATLAB:

```
F03y=sym('F03y','real');
F03_=[0, F03y, 0]; % unknown joint force
```

Newton equation of motion for the slider 3 gives

$$m_3 \, \mathbf{a}_{C_3} = \sum \mathbf{F}^{(3)} = \mathbf{F}_{23} + \mathbf{F}_{03} + \mathbf{G}_3. \tag{4.32}$$

The scalar equations of motion for the slider 3 are

$$m_3 \, a_{C_3 x} = F_{23x} + F_{03x}, \tag{4.33}$$

$$m_3 \, a_{C_3 y} = F_{23y} + F_{03y} - m_3 \, g. \tag{4.34}$$

The MATLAB commands for the equations of motion of the slider 3 are:

```
% sum F of 3
F23_  = -F32_;
eqF3_ = -m3*aC3_+F23_+F03_+G3_;
eqF3x = eqF3_(1); % (11) x-component
eqF3y = eqF3_(2); % (12) y-component
```

For the links 2 and 3 there are five equations Eqs. (4.29), (4.30), (4.31), (4.33), (4.34) with five unknowns F_{12x}, F_{12y}, F_{32x}, F_{32y}, F_{03y}. The algebraic system of five equations are solved using MATLAB:

```
sol23=solve(eqF2x,eqF2y,eqMC2z,eqF3x,eqF3y);

F12xs=eval(sol23.F12x);
F12ys=eval(sol23.F12y);
F32xs=eval(sol23.F32x);
F32ys=eval(sol23.F32y);
F03ys=eval(sol23.F03y);

F12_ = [F12xs F12ys 0];
F03_ = [0 F03ys 0];
F32_ = [F32xs F32ys 0];
```

The numerical values of the joint reaction forces are

```
% F12_ = [2024.907,-512.546,0] (N)
% F03_ = [0,-143.783,0] (N)
% F32_ = [216.373,-148.804,0] (N)
```

The free-body diagram of the driver link 1 is depicted in Fig. 4.5. The reaction force of the ground on link 1 at A is $\mathbf{F}_{01}$. The equilibrium moment that acts on link 1 is $\mathbf{M}_m$. The reaction $\mathbf{F}_{01}$ is calculated from Newton equation of motion for link 1

Fig. 4.5 Free-body diagram
for link 1

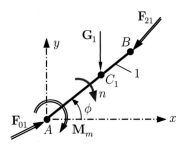

$$m_1\, a_{C_1} = \sum \mathbf{F}^{(1)} = \mathbf{F}_{01} + \mathbf{F}_{21} + \mathbf{G}_1 \implies \mathbf{F}_{01} = m_1\, a_{C_1} - \mathbf{F}_{21} - \mathbf{G}_1. \quad (4.35)$$

From Euler equation of motion for the link 1 with respect to the fixed point A the moment $\mathbf{M}_\mathrm{m}$ is calculated

$$I_A\, \alpha_1 = \sum \mathbf{M}_A^{(1)} = \mathbf{r}_{C1} \times \mathbf{G}_1 + \mathbf{r}_B \times \mathbf{F}_{21} + \mathbf{M}_\mathrm{m} \implies$$
$$\mathbf{M}_\mathrm{m} = I_A\, \alpha_1 - \mathbf{r}_{C1} \times \mathbf{G}_1 - \mathbf{r}_B \times \mathbf{F}_{21}. \quad (4.36)$$

The MATLAB commands for the joint reaction $\mathbf{F}_{01}$ and the motor moment $\mathbf{M}_\mathrm{m}$ are:

```
F21_ = -F12_;
F01_ = m1*aC1_-G1_-F21_;
IA  = IC1+m1*(AB/2)^2;
Mm_ =IA*alpha1_-cross(rC1_,G1_)-cross(rB_,F21_);
```

and the numerical results are:

```
% F01_ = [1962.873,-571.442,0] (N)
% Mm_ = [0,0,-358.628] (N m)
```

4.4 Problems

4.1 A planar mechanism is shown in Fig. 4.6. The following dimensions are given: $AB = 150\,\mathrm{mm}$, $AC = 220\,\mathrm{mm}$, $CD = 150\,\mathrm{mm}$, $DE = 200\,\mathrm{mm}$, $a = 230\,\mathrm{mm}$. Link 1 rotates at a constant angular speed $n_1 = 120$ rpm. Find the positions, velocities, accelerations, joint forces and motor (equilibrium) moment on link 1 for $\phi = 45°$.

4.2 The mechanism shown in Fig. 4.7 has the dimensions $AB = 10$ mm, $BC = 25$ mm, $CD = 17$ mm, $AD = 30$ mm, $DF = 15$ mm, $EF = 17$ mm. Link 1 rotates at a constant angular speed $n_1 = 100$ rpm. Find the positions, velocities, accelerations, joint forces and motor (equilibrium) moment on link 1 for $\phi = 30°$

.

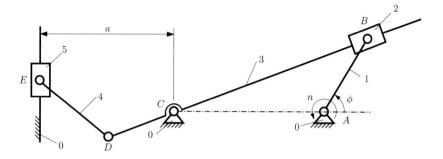

Fig. 4.6 Problem 4.1

Fig. 4.7 Problem 4.2

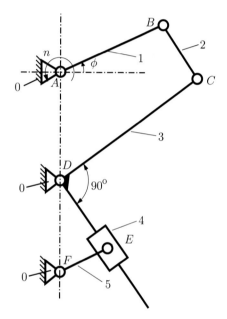

4.3 The mechanism shown in Fig. 4.8 has the dimensions $AB = 100$ mm, $AC = 280$ mm, $BD = a = 470$ mm, $DE = 220$ mm. Link 1 rotates at a constant angular speed $n_1 = -60$ rpm. Find the positions, velocities, accelerations, joint forces and motor (equilibrium) moment on link 1 for $\phi = 60°$.

4.4 The mechanism shown in Fig. 4.9 has the dimensions $AB = 80$ mm, $BC = 350$ mm, $CG = 200$ mm, $CD = 150$ mm, $a = 200$ mm, $b = 350$ mm. Link 1 rotates at a constant angular speed $n_1 = 150$ rpm. Find the positions, velocities, accelerations, joint forces and motor (equilibrium) moment on link 1 for $\phi = 120°$. Select a suitable distance from A to E.

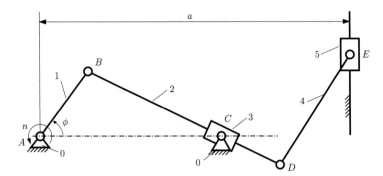

Fig. 4.8 Problem 4.3

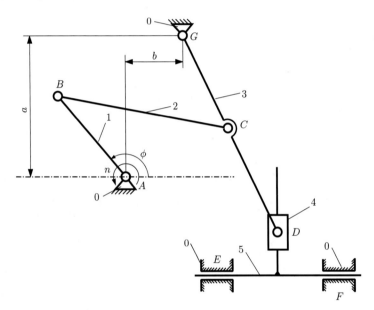

Fig. 4.9 Problem 4.4

4.5 The mechanism shown in Fig. 4.10 has the dimensions $AB = 150$ mm, $BC =$ 300 mm, $BE = 600$ mm, $CE = 850$ mm, $CD = 330$ mm, $EF = 1200$ mm, $a = 350$ mm, $b = 200$ mm, $c = 100$ mm. Link 1 rotates at a constant angular speed $n_1 = 200$ rpm. Find the positions, velocities, accelerations, joint forces and motor (equilibrium) moment on link 1 for $\phi = 120°$.

Fig. 4.10 Problem 4.5

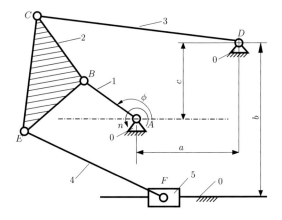

References

1. E.A. Avallone, T. Baumeister, A. Sadegh, *Marks' Standard Handbook for Mechanical Engineers*, 11th edn. (McGraw-Hill Education, New York, 2007)
2. I.I. Artobolevski, *Mechanisms in Modern Engineering Design* (MIR, Moscow, 1977)
3. M. Atanasiu, *Mechanics (Mecanica)* (EDP, Bucharest, 1973)
4. H. Baruh, *Analytical Dynamics* (WCB/McGraw-Hill, Boston, 1999)
5. A. Bedford, W. Fowler, *Dynamics* (Addison Wesley, Menlo Park, CA, 1999)
6. A. Bedford, W. Fowler, *Statics* (Addison Wesley, Menlo Park, CA, 1999)
7. F.P. Beer et al., *Vector Mechanics for Engineers: Statics and Dynamics* (McGraw-Hill, New York, NY, 2016)
8. M. Buculei, D. Bagnaru, G. Nanu, D.B. Marghitu, *Computing Methods in the Analysis of the Mechanisms with Bars* (Scrisul Romanesc, Craiova, 1986)
9. M.I. Buculei, *Mechanisms* (University of Craiova Press, Craiova, Romania, 1976)
10. D. Bolcu, S. Rizescu, *Mecanica* (EDP, Bucharest, 2001)
11. J. Billingsley, *Essential of Dynamics and Vibration* (Springer, 2018)
12. R. Budynas, K.J. Nisbett, *Shigley's Mechanical Engineering Design*, 9th edn. (McGraw-Hill, New York, 2013)
13. J..A. Collins, H.R. Busby, G.H. Staab, *Mechanical Design of Machine Elements and Machines*, 2nd edn. (Wiley, 2009)
14. M. Crespo da Silva, *Intermediate Dynamics for Engineers* (McGraw-Hill, New York, 2004)
15. A.G. Erdman, G.N. Sandor, *Mechanisms Design* (Prentice-Hall, Upper Saddle River, NJ, 1984)
16. M. Dupac, D.B. Marghitu, *Engineering Applications: Analytical and Numerical Calculation with MATLAB* (Wiley, Hoboken, NJ, 2021)
17. A. Ertas, J.C. Jones, *The Engineering Design Process* (Wiley, New York, 1996)
18. F. Freudenstein, An application of Boolean algebra to the motion of epicyclic drivers. J. Eng. Ind. pp. 176–182 (1971)
19. J.H. Ginsberg, *Advanced Engineering Dynamics* (Cambridge University Press, Cambridge, 1995)
20. H. Goldstein, *Classical Mechanics* (Addison-Wesley, Redwood City, CA, 1989)
21. D.T. Greenwood, *Principles of Dynamics* (Prentice-Hall, Englewood Cliffs, NJ, 1998)
22. A.S. Hall, A.R. Holowenko, H.G. Laughlin, *Schaum's Outline of Machine Design* (McGraw-Hill, New York, 2013)
23. B.G. Hamrock, B. Jacobson, S.R. Schmid, *Fundamentals of Machine Elements* (McGraw-Hill, New York, 1999)
24. R.C. Hibbeler, *Engineering Mechanics: Dynamics* (Prentice Hall, 2010)

25. T..E.. Honein, O..M.. O'Reilly, On the Gibbs–Appell equations for the dynamics of rigid bodies. J. Appl. Mech. **88**, 074501-1 (2021)
26. R.C. Juvinall, K.M. Marshek, *Fundamentals of Machine Component Design*, 5th edn. (Wiley, New York, 2010)
27. T.R. Kane, *Analytical Elements of Mechanics*, vol. 1 (Academic Press, New York, 1959)
28. T.R. Kane, *Analytical Elements of Mechanics*, vol. 2 (Academic Press, New York, 1961)
29. T.R. Kane, D.A. Levinson, The use of Kane's dynamical equations in robotics. MIT Int. J. Robot. Res. **3**, 3–21 (1983)
30. T.R. Kane, P.W. Likins, D.A. Levinson, *Spacecraft Dynamics* (McGraw-Hill, New York, 1983)
31. T.R. Kane, D.A. Levinson, *Dynamics* (McGraw-Hill, New York, 1985)
32. K. Lingaiah, *Machine Design Databook*, 2nd edn. (McGraw-Hill Education, New York, 2003)
33. N.I. Manolescu, F. Kovacs, A. Oranescu, *The Theory of Mechanisms and Machines* (EDP, Bucharest, 1972)
34. D.B. Marghitu, *Mechanical Engineer's Handbook* (Academic Press, San Diego, CA, 2001)
35. D.B. Marghitu, M.J. Crocker, *Analytical Elements of Mechanisms* (Cambridge University Press, Cambridge, 2001)
36. D.B. Marghitu, *Kinematic Chains and Machine Component Design* (Elsevier, Amsterdam, 2005)
37. D.B. Marghitu, *Mechanisms and Robots Analysis with MATLAB* (Springer, New York, N.Y., 2009)
38. D.B. Marghitu, M. Dupac, *Advanced Dynamics: Analytical and Numerical Calculations with MATLAB* (Springer, New York, N.Y., 2012)
39. D.B. Marghitu, M. Dupac, H.M. Nels, *Statics with MATLAB* (Springer, New York, N.Y., 2013)
40. D.B. Marghitu, D. Cojocaru, *Advances in Robot Design and Intelligent Control* (Springer International Publishing, Cham, Switzerland, 2016), pp. 317–325
41. D.J. McGill, W.W. King, *Engineering Mechanics: Statics and an Introduction to Dynamics* (PWS Publishing Company, Boston, 1995)
42. J.L. Meriam, L.G. Kraige, *Engineering Mechanics: Dynamics* (Wiley, New York, 2007)
43. C.R. Mischke, Prediction of stochastic endurance strength. Trans. ASME, J. Vib. Acoust. Stress Reliab. Des. **109**(1), 113–122 (1987)
44. L. Meirovitch, *Methods of Analytical Dynamics* (Dover, 2003)
45. R.L. Mott, *Machine Elements in Mechanical Design* (Prentice Hall, Upper Saddle River, NJ, 1999)
46. W.A. Nash, *Strength of Materials*. Schaum's Outline Series (McGraw-Hill, New York, 1972)
47. R.L. Norton, *Machine Design* (Prentice-Hall, Upper Saddle River, NJ, 1996)
48. R.L. Norton, *Design of Machinery* (McGraw-Hill, New York, 1999)
49. O.M. O'Reilly, *Intermediate Dynamics for Engineers Newton-Euler and Lagrangian Mechanics* (Cambridge University Press, UK, 2020)
50. O.M. O'Reilly, *Engineering Dynamics: A Primer* (Springer, NY, 2010)
51. W.C. Orthwein, *Machine Component Design* (West Publishing Company, St. Paul, 1990)
52. L.A. Pars, *A Treatise on Analytical Dynamics* (Wiley, New York, 1965)
53. F. Reuleaux, *The Kinematics of Machinery* (Dover, New York, 1963)
54. D. Planchard, M. Planchard, *SolidWorks 2013 Tutorial with Video Instruction* (SDC Publications, 2013)
55. I. Popescu, *Mechanisms* (University of Craiova Press, Craiova, Romania, 1990)
56. I. Popescu, C. Ungureanu, *Structural Synthesis and Kinematics of Mechanisms with Bars* (Universitaria Press, Craiova, Romania, 2000)
57. I. Popescu, L. Luca, M. Cherciu, D.B. Marghitu, *Mechanisms for Generating Mathematical Curves* (Springer Nature, Switzerland, 2020)
58. C.A. Rubin, *The Student Edition of Working Model* (Addison-Wesley Publishing Company, Reading, MA, 1995)
59. J. Ragan, D.B. Marghitu, Impact of a kinematic link with MATLAB and solidworks. Appl. Mech. Mater. **430**, 170–177 (2013)

60. J. Ragan, D.B. Marghitu, MATLAB dynamics of a free link with elastic impact, in *International Conference on Mechanical Engineering, ICOME 2013*, May 16–17, 2013, Craiova, Romania (2013)
61. J.C. Samin, P. Fisette, *Symbolic Modeling of Multibody Systems* (Kluwer, 2003)
62. A.A. Shabana, *Computational Dynamics* (Wiley, New York, 2010)
63. I.H. Shames, *Engineering Mechanics - Statics and Dynamics* (Prentice-Hall, Upper Saddle River, NJ, 1997)
64. J.E. Shigley, C.R. Mischke, *Mechanical Engineering Design* (McGraw-Hill, New York, 1989)
65. J.E. Shigley, C.R. Mischke, R.G. Budynas, *Mechanical Engineering Design*, 7th edn. (McGraw-Hill, New York, 2004)
66. J.E. Shigley, J.J. Uicker, *Theory of Machines and Mechanisms* (McGraw-Hill, New York, 1995)
67. D. Smith, *Engineering Computation with MATLAB* (Pearson Education, Upper Saddle River, NJ, 2008)
68. R.W. Soutas-Little, D.J. Inman, *Engineering Mechanics: Statics and Dynamics* (Prentice-Hall, Upper Saddle River, NJ, 1999)
69. J. Sticklen, M.T. Eskil, *An Introduction to Technical Problem Solving with MATLAB* (Great Lakes Press, Wildwood, MO, 2006)
70. A.C. Ugural, *Mechanical Design* (McGraw-Hill, New York, 2004)
71. R. Voinea, D. Voiculescu, V. Ceausu, *Mechanics (Mecanica)* (EDP, Bucharest, 1983)
72. K.J. Waldron, G.L. Kinzel, *Kinematics, Dynamics, and Design of Machinery* (Wiley, New York, 1999)
73. J.H. Williams Jr., *Fundamentals of Applied Dynamics* (Wiley, New York, 1996)
74. C.E. Wilson, J.P. Sadler, *Kinematics and Dynamics of Machinery* (Harper Collins College Publishers, New York, 1991)
75. H.B. Wilson, L.H. Turcotte, D. Halpern, *Advanced Mathematics and Mechanics Applications Using MATLAB* (Chapman & Hall/CRC, 2003)
76. S. Wolfram, *Mathematica* (Wolfram Media/Cambridge University Press, Cambridge, 1999)
77. National Council of Examiners for Engineering and Surveying (NCEES), *Fundamentals of Engineering. Supplied-Reference Handbook*, Clemson, SC (2001)
78. * * * ,*The Theory of Mechanisms and Machines* (Teoria mehanizmov i masin), Vassaia scola, Minsk, Russia (1970)
79. eCourses - University of Oklahoma: http://ecourses.ou.edu/home.htm
80. https://www.mathworks.com
81. http://www.eng.auburn.edu/~marghitu/
82. https://www.solidworks.com
83. https://www.wolfram.com

Chapter 5
Contour Analysis of a Mechanism with Two Dyads

Abstract A planar R-RRT-RTR mechanism with two dyads is considered. There are two independent closed contours for the velocity and acceleration fields. For dynamic analysis D'Alembert principle is employed to calculate individual joint forces.

The planar R-RRT-RTR mechanism analyzed in Chap. 4 is depicted in Fig. 5.1a. Figure 5.1a shows the contour diagram and the two independent loops are shown in Fig. 5.1b, c. The input data and the position data are:

```
AB = 0.20 ; % m
BC = 0.21 ; % m
CD = 0.39 ; % m
CF = 0.45 ; % m
a  = 0.30 ; % m
b  = 0.25 ; % m
c  = 0.05 ; % m
phi = 45*pi/180 ; % rad
n = -500 ;          % rpm
% phi = phi1 = 45 (degrees)
% rA_ = [0, 0, 0] (m)
% rE_ = [ 0.300, -0.250, 0] (m)
% rB_ = [ 0.141,  0.141, 0] (m)
% rD_ = [ 0.734, 0.05, 0] (m)
% phi2 = -8.764 (degrees)
% rC_ = [ 0.349,  0.109, 0] (m)
% phi4 = phi5 = 82.242 (degrees)
% rF_ = [ 0.288, -0.336, 0] (m)
```

D. B. Marghitu et al., *Mechanical Simulation with MATLAB®*,
Springer Tracts in Mechanical Engineering,
https://doi.org/10.1007/978-3-030-88102-3_5

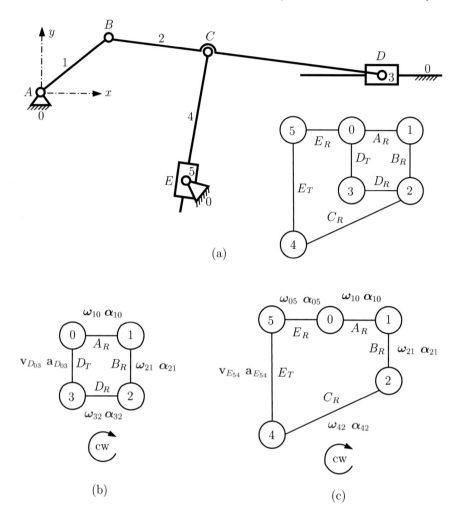

(a)

(b)

(c)

Fig. 5.1 R-RRT-RTR mechanism and contour diagrams

5.1 Velocity and Acceleration Analysis

The first contour, shown in Fig. 5.1b, is formed by the links: 0–1–2–3–0. A clockwise direction is selected. The closed loop equations for velocities are

$$\boldsymbol{\omega}_{10} + \boldsymbol{\omega}_{21} + \boldsymbol{\omega}_{32} = \mathbf{0},$$

$$\mathbf{r}_B \times \boldsymbol{\omega}_{21} + \mathbf{r}_D \times \boldsymbol{\omega}_{32} + \mathbf{v}_{D_{03}}^{\mathrm{r}} = \mathbf{0}. \tag{5.1}$$

The unknowns are

$$\boldsymbol{\omega}_{21} = \omega_{21}\,\mathbf{k},\;\; \boldsymbol{\omega}_{32} = \omega_{32}\,\mathbf{k},\;\; \text{and}\;\; \mathbf{v}^r_{D_{03}} = \mathbf{v}_{03} = v_{03}\,\mathbf{I}.$$

The MATLAB commands for solving the previous system of algebraic equations are:

```
n = -500; % (rpm) angular speed of 1
omega1_ = [0 0 pi*n/30]; % angular velocity of 1
alpha1_ = [0 0 0]; % angular acceleration of 1

omega10_ = omega1_;
syms omega21z omega32z v03
omega21v_=[0 0 omega21z];% angular velocity of 2 relative to 1
omega32v_=[0 0 omega32z];% angular velocity of 3 relative to 2
v03v_ =[v03 0 0]; % velocity of D on 0 relative to D on 3
% First contour velocity equations
eqIomega_ = omega10_ + omega21v_ + omega32v_;
eqIvz = eqIomega_(3); % (1)
eqIv = cross(rB_, omega21v_) + cross(rD_, omega32v_) + v03v_;
eqIvx = eqIv(1); % (2)
eqIvy = eqIv(2); % (3)
% Eqs.(1)(2)(3) =>
solIv=solve(eqIvz,eqIvx,eqIvy);
omega21_ = [ 0 0 eval(solIv.omega21z) ];
omega32_ = [ 0 0 eval(solIv.omega32z) ];
vD03 = eval(solIv.v03);
vD03_ = [vD03 0 0];
```

The relative velocity results are:

```
% omega21_ = [ 0, 0, 64.847 ] (rad/s)
% omega32_ = [ 0, 0, -12.487 ] (rad/s)
% vD03 = -8.546 (m/s)
% vD03_ = [ -8.546, 0, 0 ] (m/s)
```

The absolute angular velocity of link 2 is:

```
omega20_ = omega10_+omega21_; % angular velocity of 2
fprintf('omega20_= [%d, %d, %6.3f] (rad/s)\n',omega20_)
% omega20_= [0, 0, 12.487] (rad/s)
```

and the velocity of joint D is:

```
vD3_ = -vD03_; % velocity of D
vD_ = vD3_;
fprintf('vD3_ = [ %6.3f, %d, %d ] (m/s)\n', vD3_)
% vD3_ = [  8.546, 0, 0 ] (m/s)
```

The equations for the acceleration field are

$$\boldsymbol{\alpha}_{10} + \boldsymbol{\alpha}_{21} + \boldsymbol{\alpha}_{32} = \mathbf{0},$$

$$\mathbf{r}_B \times \boldsymbol{\alpha}_{21} + \mathbf{r}_D \times \boldsymbol{\alpha}_{32} + \mathbf{a}_{D_{03}}^r - \omega_{10}^2 \mathbf{r}_B - \omega_{20}^2 \mathbf{r}_{BD} = \mathbf{0}, \qquad (5.2)$$

with three unknowns:

$$\boldsymbol{\alpha}_{21} = \alpha_{21}\,\mathbf{k}, \quad \boldsymbol{\alpha}_{32} = \alpha_{32}\,\mathbf{k}, \quad \text{and } \mathbf{a}_{D_{03}}^r = \mathbf{a}_{03} = a_{03}\,\mathbf{1}.$$

The system of equations is solved with MATLAB:

```
alpha10_ = alpha1_;
syms alpha21z alpha32z a03
alpha21v_=[0 0 alpha21z];% angular acceleration of 2 relative to 1
alpha32v_=[0 0 alpha32z];% angular acceleration of 3 relative to 2
a03v_=[a03 0 0];% acceleration of D on 0 relative to D on 3
% First contour acceleration equations
eqIalpha_ = alpha10_ + alpha21v_ + alpha32v_;
eqIaz = eqIalpha_(3); % (4)
eqIa_=cross(rB_,alpha21v_)+cross(rD_,alpha32v_)+a03v_...
-dot(omega1_,omega1_)*rB_-dot(omega20_,omega20_)*(rD_-rB_);
eqIax = eqIa_(1); % (5)
eqIay = eqIa_(2); % (6)
% Eqs.(4)(5)(6) =>
solIa=solve(eqIaz,eqIax,eqIay);
alpha21_ = [ 0 0 eval(solIa.alpha21z) ];
alpha32_ = [ 0 0 eval(solIa.alpha32z) ];
aD03_ = [ eval(solIa.a03) 0 0];
```

and the results are:

```
fprintf('alpha21_ = [ %g, %g, %6.3f ] (rad/s^2)\n', alpha21_)
fprintf('alpha32_ = [ %g, %g, %6.3f ] (rad/s^2)\n', alpha32_)
fprintf('aD03 = %6.3f (m/s^2)\n', eval(solIa.a03))
fprintf('aD03_ = [ %6.3f, %d, %d ] (m/s^2)\n', aD03_)
fprintf('\n')
% alpha21_ = [ 0, 0, 629.786 ] (rad/s^2)
% alpha32_ = [ 0, 0, -629.786 ] (rad/s^2)
% aD03 = 422.604 (m/s^2)
% aD03_ = [ 422.604, 0, 0 ] (m/s^2)
```

The absolute accelerations are:

```
alpha20_ = alpha10_+alpha21_; % angular acceleration of 2
aD3_ = -aD03_; % acceleration of D
aD_ = aD3_;
fprintf('alpha20_= [%d, %d, %6.3f] (rad/s^2)\n',alpha20_)
fprintf('aD3_ = [ %6.3f, %d, %d ] (m/s^2)\n', aD3_)
fprintf('\n')
% alpha20_= [0, 0, 629.786] (rad/s^2)
% aD3_ = [ -422.604, 0, 0 ] (m/s^2)
```

The second closed loop is selected as 0–1–2–4–5–0 as seen in Fig. 5.1c. The velocity equations, for the clockwise path, are

$$\boldsymbol{\omega}_{10} + \boldsymbol{\omega}_{21} + \boldsymbol{\omega}_{42} + \boldsymbol{\omega}_{05} = \mathbf{0},$$
$$\mathbf{r}_B \times \boldsymbol{\omega}_{21} + \mathbf{r}_C \times \boldsymbol{\omega}_{42} + \mathbf{r}_E \times \boldsymbol{\omega}_{05} + \mathbf{v}^r_{E_{54}} = \mathbf{0}. \tag{5.3}$$

The unknown relative velocities are:

$$\omega_{42} = \omega_{42}\,\mathbf{k}, \quad \boldsymbol{\omega}_{05} = \omega_{05}\,\mathbf{k}, \quad \text{and } \mathbf{v}^r_{E_{54}} = \mathbf{v}_{E_{54}} = v_{E_{54}}(\cos\phi_4\,\mathbf{1} + \sin\phi_4\,\mathbf{J}).$$

The MATLAB code for velocity equations is:

```
syms omega42z vE54 omega05z
omega42v_=[0 0 omega42z];% angular velocity of 4 relative to 2
omega05v_=[0 0 omega05z];% angular velocity of 0 relative to 5
vE54v_=vE54*[ cos(phi4) sin(phi4) 0];% velocity of E on 5 relative to 4

eqIIomega_ = omega10_ + omega21_ + omega42v_ + omega05v_;
eqIIvz = eqIIomega_(3); % (7)
eqIIv = cross(rB_,omega21_) + cross(rC_,omega42v_) ...
        + cross(rE_,omega05v_) + vE54v_;
eqIIvx = eqIIv(1); % (8)
eqIIvy = eqIIv(2); % (9)
```

The solution for the relative velocities is:

```
% Eqs.(7)(8)(9) =>
solIIv=solve(eqIIvz,eqIIvx,eqIIvy);
omega42_ = [ 0 0 eval(solIIv.omega42z) ];
omega05_ = [ 0 0 eval(solIIv.omega05z) ];
vE54_ = eval(solIIv.vE54)*[ cos(phi4) sin(phi4) 0];

fprintf('omega42_ = [ %g, %g, %6.3f ] (rad/s)\n', omega42_)
fprintf('omega05_ = [ %g, %g, %6.3f ] (rad/s)\n', omega05_)
fprintf('vE54 = %6.3f (m/s)\n', eval(solIIv.vE54))
fprintf('vE54_ = [ %6.3f, %6.3f, %d ] (m/s)\n', vE54_)
```

```
fprintf('\n')
```

```
% omega42_ = [ 0, 0, -35.596 ] (rad/s)
% omega05_ = [ 0, 0,  23.109 ] (rad/s)
% vE54 =  3.716 (m/s)
% vE54_ = [  0.502,   3.681, 0 ] (m/s)
```

The absolute angular velocity of link 4 is:

```
omega40_ = omega20_+omega42_;
fprintf('omega40_=omega50_= [%d, %d, %6.3f] (rad/s)\n',omega40_)
% omega40_=omega50_= [0, 0, -23.109] (rad/s)
```

The acceleration equations for the second contour are:

$$\boldsymbol{\alpha}_{10} + \boldsymbol{\alpha}_{21} + \boldsymbol{\alpha}_{42} + \boldsymbol{\alpha}_{05} = \mathbf{0},$$
$$\mathbf{r}_B \times \boldsymbol{\alpha}_{21} + \mathbf{r}_C \times \boldsymbol{\alpha}_{42} + \mathbf{r}_E \times \boldsymbol{\alpha}_{05}$$
$$+ \mathbf{a}^{\mathrm{r}}_{E_{54}} + \mathbf{a}^{\mathrm{c}}_{E_{54}}$$
$$- \omega_{10}^2 \mathbf{r}_B - \omega_{20}^2 \mathbf{r}_{BC} - \omega_{40}^2 \mathbf{r}_{CE} = \mathbf{0}, \tag{5.4}$$

where

$$\boldsymbol{\alpha}_{42} = \alpha_{42}\,\mathbf{k}, \quad \boldsymbol{\alpha}_{05} = \alpha_{05}\,\mathbf{k},$$
$$\mathbf{a}^{\mathrm{r}}_{E_{54}} = \mathbf{a}_{E_{54}} = a_{E_{54}}\,\cos\phi_4\,\mathbf{1} + a_{E_{54}}\,\sin\phi_4\,\mathbf{J},$$
$$\mathbf{a}^{\mathrm{c}}_{E_{54}} = \mathbf{a}^{\mathrm{c}}_{E_{54}} = 2\,\boldsymbol{\omega}_{40} \times \mathbf{v}_{E_{54}}.$$

The MATLAB commands for acceleration analysis are:

```
syms alpha42z alpha05z aE54

alpha42v_=[0 0 alpha42z];% angular acceleration of 4 relative to 2
alpha05v_=[0 0 alpha05z];% angular acceleration of 0 relative to 5
aE54v_=aE54*[cos(phi4) sin(phi4) 0];% acceleration of E on 5 relative to 4

eqIIalpha_ = alpha10_ + alpha21_ + alpha42v_ + alpha05v_;
eqIIaz = eqIIalpha_(3); % (10)
eqIIa_=...
cross(rB_,alpha21_)+cross(rC_,alpha42v_)+cross(rE_,alpha05v_)...
+aE54v_+2*cross(omega40_,vE54_)...
-dot(omega10_,omega10_)*rB_-dot(omega20_,omega20_)*(rC_-rB_)...
-dot(omega40_,omega40_)*(rE_-rC_);
eqIIax=eqIIa_(1); % (11)
eqIIay=eqIIa_(2); % (12)
```

The solution of the acceleration equations is:

```
% Eqs.(10)(11)(12) =>
solIIa=solve(eqIIaz,eqIIax,eqIIay);
alpha42_ = [ 0 0 eval(solIIa.alpha42z) ];
alpha05_ = [ 0 0 eval(solIIa.alpha05z) ] ;
aE54_ = eval(solIIa.aE54)*[ cos(phi4) sin(phi4) 0];


fprintf('alpha42_ = [ %g, %g, %6.3f ] (rad/s^2)\n', alpha42_)
fprintf('alpha05_ = [ %g, %g, %6.3f ] (rad/s^2)\n', alpha05_)
fprintf('aE54 = %6.3f (m/s^2)\n', eval(solIIa.aE54))
fprintf('aE54_ = [ %6.3f, %6.3f, %d ] (m/s^2)\n', aE54_)
fprintf('\n')


% alpha42_ = [ 0, 0, -104.567 ] (rad/s^2)
% alpha05_ = [ 0, 0, -525.219 ] (rad/s^2)
% aE54 = 109.981 (m/s^2)
% aE54_ = [ 14.847, 108.974, 0 ] (m/s^2)
```

The absolute acceleration of links 4 and 5 is:

```
alpha50_ = - alpha05_;
alpha40_ = alpha50_;
fprintf('alpha40_=alpha50_=[%d, %d, %6.3f] (rad/s^2)\n',alpha50_)
% alpha40_=alpha50_=[0, 0, 525.219] (rad/s^2)
```

5.2 Contour Dynamic Force Analysis with D'Alembert Principle

5.2.1 Reaction Force F_{24}

The revolute joint at C between link 2 and the link 4 is replaced with the joint reaction force $\mathbf{F}_{24}$ as shown in Fig. 5.2. The joint reaction force of the link 2 on the link 4, $\mathbf{F}_{24}$, acts at C and has two unknown components $\mathbf{F}_{24} = F_{24x}\mathbf{i} + F_{24y}\mathbf{j}$.

The sum of the forces on link 4 is equal to zero, Fig. 5.2a, or

$$\sum \mathbf{F}^{(4)} = \mathbf{F}_{in\,4} + \mathbf{G}_4 + \mathbf{F}_{24} + \mathbf{F}_{54} = 0, \qquad (5.5)$$

where $\mathbf{F}_{54}$ is the reaction force of the slider 5 on the link 4 and is perpendicular to the sliding direction CE, i.e., $\mathbf{F}_{54} \cdot \mathbf{r}_{CE} = 0$. Equation (5.5) is scalar multiplied with $\mathbf{r}_{CE}$ and the following equation is obtained

Fig. 5.2 Force diagram for calculating $\mathbf{F}_{24}$

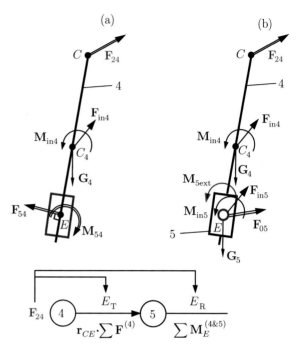

$$\mathbf{r}_{CE} \cdot \sum \mathbf{F}^{(4)} = (\mathbf{F}_{in\,4} + \mathbf{G}_4 + \mathbf{F}_{24}) \cdot \mathbf{r}_{CE} = 0. \tag{5.6}$$

Equation (5.6) does not contain the reaction $\mathbf{F}_{54}$ because $\mathbf{F}_{54} \cdot \mathbf{r}_{CE} = 0$. Next for the links 4 and 5 the sum of the moments with respect to the revolute joint E, Fig. 5.2b, gives

$$\sum \mathbf{M}_E^{(4\&5)} = \mathbf{r}_{EC} \times \mathbf{F}_{24} + \mathbf{r}_{EC_4} \times (\mathbf{F}_{in\,4} + \mathbf{G}_4) + \mathbf{M}_{in\,4} + \mathbf{M}_{in\,5} + \mathbf{M}_e = \mathbf{0}, \tag{5.7}$$

where the external moment on slider 5 is $\mathbf{M}_e = \mathbf{M}_{5ext}$. Equation (5.7) has only a component on z−axis. Equations (5.6) and (5.7) represent a system of two equations with two unknowns F_{24x} and F_{24y}. The MATLAB commands for Eqs. (5.6) and (5.7) are:

```
F24x=sym('F24x','real');
F24y=sym('F24y','real');
F24_=[F24x, F24y, 0]; % unknown joint force
% E_P: (sumF4).CE = 0
% (Fin4_+G4_+F24_).(rE_-rC_)=0
SF4CE = dot(Fin4_+G4_+F24_,rE_-rC_);
% E_R: sumM45E_ = 0_
```

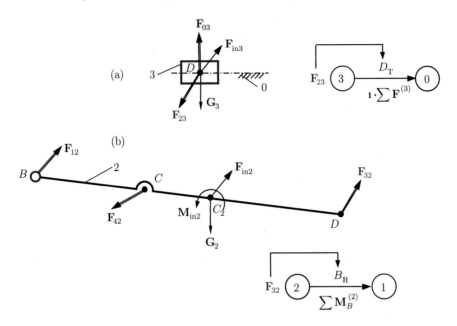

Fig. 5.3 Force diagram for calculating $\mathbf{F}_{23}$

```
% EC_xF24_+EC4x(Fin4_+G4_)+Min4_+Min5_+Me_=0
sumM45E_ = cross(rC_-rE_,F24_)+cross(rC4_-rE_,Fin4_+G4_)...
           +Min4_+Min5_+Me_;
sumM45Ez = sumM45E_(3);
```

The systems of equations are solved with MATLAB:

```
solF24=solve(SF4CE,sumM45Ez);
F24xs=eval(solF24.F24x);
F24ys=eval(solF24.F24y);
F24_ = [F24xs F24ys 0];
F42_ = -F24_;
% F24_ = [2630.233,-484.662,0] (N)
```

5.2.2 Reaction Force $\mathbf{F}_{23}$

The revolute joint at D between link 2 and the link 3 is replaced with the joint reaction force $\mathbf{F}_{23}$ as shown in Fig. 5.3. The joint reaction force of the link 2 on the link 3, $\mathbf{F}_{23}$, acts at D and has two unknown components $\mathbf{F}_{23} = F_{23x} \mathbf{I} + F_{23y} \mathbf{J}$.
   The sum of the forces on link 3 is equal to zero, Fig. 5.3a, or

$$\sum \mathbf{F}^{(3)} = \mathbf{F}_{\text{in}\,3} + \mathbf{G}_3 + \mathbf{F}_{23} + \mathbf{F}_{03} = \mathbf{0}, \tag{5.8}$$

where $\mathbf{F}_{03}$ is the reaction force of the ground on slider 3 and is perpendicular to the sliding direction $x-$axis, i.e., $\mathbf{F}_{03} \cdot \mathbf{i} = 0$. Equation (5.8) is scalar multiplied with $\mathbf{i}$ and the following relation is obtained

$$\left(\sum \mathbf{F}^{(3)} \right) \cdot \mathbf{i} = (\mathbf{F}_{\text{in}\,3} + \mathbf{G}_3 + \mathbf{F}_{23}) \cdot \mathbf{i} = 0. \tag{5.9}$$

Next for the link 2, Fig. 5.3b, the sum of the moments with respect to the revolute joint B gives

$$\sum \mathbf{M}_B^{(2)} =$$
$$\mathbf{r}_{BD} \times (-\mathbf{F}_{23}) + \mathbf{r}_{BC} \times (-\mathbf{F}_{24}) + \mathbf{r}_{BC_2} \times (\mathbf{F}_{\text{in}\,2} + \mathbf{G}_2) + \mathbf{M}_{\text{in}\,2} = \mathbf{0}. \tag{5.10}$$

From Eqs. (5.9) and (5.10) the two unknowns F_{23x} and F_{23y} are calculated. The MATLAB commands for the reaction $\mathbf{F}_{23}$ are:

```
F23x=sym('F23x','real');
F23y=sym('F23y','real');
F23_=[F23x, F23y, 0]; % unknown joint force
% D_P: sumF3_(1) = 0
% (Fin3_+G3_+F23_)(1) = 0
SF3_ = Fin3_+G3_+F23_;
SF3x = SF3_(1);
% B_R: sumM2B_ = 0_
% BDx(-F23_)+BCx(-F24_)+BC2x(Fin2_+G2_)+Min2_ = 0_
sumM2B_ = cross(rD_-rB_,-F23_)+cross(rC_-rB_,-F24_)...
          +cross(rC2_-rB_,Fin2_+G2_)+Min2_;
sumM2Bz = sumM2B_(3);
solF23=solve(SF3x,sumM2Bz);
F23xs=eval(solF23.F23x);
F23ys=eval(solF23.F23y);
F23_ = [F23xs F23ys 0];
% F23_ = [-216.373,148.804,0] (N)
```

s

5.2.3 Reaction Force $\mathbf{F}_{12}$

The revolute joint at B between link 1 and the link 2 is replaced with the joint reaction force $\mathbf{F}_{12}$ as shown in Fig. 5.4. The joint reaction force of the link 1 on the link 2, $\mathbf{F}_{12}$, acts at B and has two unknown components $\mathbf{F}_{12} = F_{12x}\,\mathbf{i} + F_{12y}\,\mathbf{J}$.

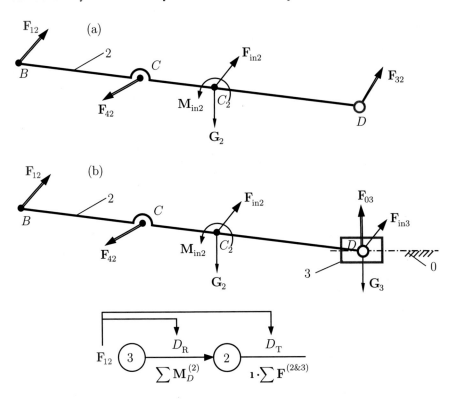

Fig. 5.4 Force diagram for calculating $\mathbf{F}_{12}$

For the link 2, Fig. 5.4a, the sum of the moments with respect to the revolute joint D is zero

$$\sum \mathbf{M}_D^{(2)} =$$
$$\mathbf{r}_{DB} \times \mathbf{F}_{12} + \mathbf{r}_{DC} \times (-\mathbf{F}_{24}) + \mathbf{r}_{DC_2} \times (\mathbf{F}_{\text{in}2} + \mathbf{G}_2) + \mathbf{M}_{\text{in}2} = \mathbf{0}. \quad (5.11)$$

The sum of the forces on the links 2 and 3 is equal to zero, Fig. 5.4b, or

$$\sum \mathbf{F}^{(2\&3)} = \mathbf{F}_{12} - \mathbf{F}_{24} + \mathbf{F}_{\text{in}2} + \mathbf{G}_2 + \mathbf{F}_{\text{in}3} + \mathbf{G}_3 + \mathbf{F}_{03} = \mathbf{0}, \quad (5.12)$$

where $\mathbf{F}_{03}$ is the reaction force of the ground on slider 3 and is perpendicular to the sliding direction $x-$axis, i.e., $\mathbf{F}_{03} \cdot \mathbf{1} = 0$. Equation (5.12) is scalar multiplied with $\mathbf{1}$ and the following relation is obtained

$$\left(\sum \mathbf{F}^{(2\&3)}\right) \cdot \mathbf{1} = (\mathbf{F}_{12} - \mathbf{F}_{24} + \mathbf{F}_{\text{in}2} + \mathbf{G}_2 + \mathbf{F}_{\text{in}3} + \mathbf{G}_3) \cdot \mathbf{1} = 0. \quad (5.13)$$

From Eqs. (5.11) and (5.13) the two unknowns F_{12x} and F_{12y} are calculated and the MATLAB commands for the reaction $\mathbf{F}_{12}$ are:

```
F12x=sym('F12x','real');
F12y=sym('F12y','real');
F12_=[F12x, F12y, 0]; % unknown joint force
% D_R: sumM2D_ = 0_
% DB_xF12_+DC_x(-F24_)+DC2_x(Fin2_+G2_)+Min2_ = 0_
sumM2D_ = cross(rB_-rD_,F12_)+cross(rC_-rD_,-F24_)...
          +cross(rC2_-rD_,Fin2_+G2_)+Min2_;
sumM2Dz = sumM2D_(3);
% D_P: sumF23_(1) = 0
% (F12_+(-F24)+Fin2_+G2_+Fin3_+G3_)(1) = 0
SF23_ = F12_+(-F24_)+Fin2_+G2_+Fin3_+G3_;
SF23x = SF23_(1);
solF12=solve(sumM2Dz,SF23x);
F12xs=eval(solF12.F12x);
F12ys=eval(solF12.F12y);
F12_ = [F12xs F12ys 0];
F21_ = -F12_;
% F12_ = [2024.907,-512.546,0] (N)
```

5.2.4   Reaction Force $\mathbf{F}_{03}$

The prismatic joint at D between link 0 and the slider 3 is replaced with the joint reaction force $\mathbf{F}_{03}$ as shown in Fig. 5.5. The joint reaction force of the link 0 on the slider 3, $\mathbf{F}_{03}$, acts at D and is perpendicular to the sliding direction $x-$axis, i.e., $\mathbf{F}_{03} = F_{03y} \mathbf{J}$. The application point of joint reaction force $\mathbf{F}_{03}$ acts at D because sum of the forces on link about D is zero, $\sum \mathbf{M}_D^{(3)} = \mathbf{0}$, as shown in Fig. 5.5a. For the links 3 and 2 the sum of the moments with respect to the revolute joint at B, Fig. 5.5b, gives

$$\sum \mathbf{M}_B^{(3\&2)} = \mathbf{r}_{BD} \times (\mathbf{F}_{03} + \mathbf{F}_{in\,3} + \mathbf{G}_3) + \mathbf{r}_{BC} \times (-\mathbf{F}_{24})$$
$$+\mathbf{r}_{BC_2} \times (\mathbf{F}_{in\,2} + \mathbf{G}_2) + \mathbf{M}_{in\,2} = \mathbf{0}. \tag{5.14}$$

Equation (5.14) has only a component on $z-$axis and can be solved in terms of F_{03y}. The MATLAB commands for the reaction $\mathbf{F}_{03}$ are:

```
F03y=sym('F03y','real');
F03_=[0, F03y, 0]; % unknown joint force
% B_R: sumM32B_ = 0_
% BD_x(F03_+Fin3_+G3_)+BC_x(-F24_)+BC2_x(Fin2_+G2_)+Min2_ = 0_
```

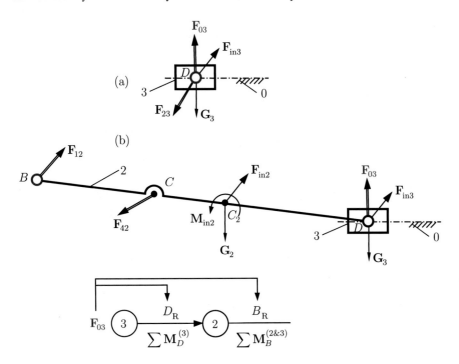

Fig. 5.5 Force diagram for calculating $\mathbf{F}_{03}$

```
sumM32B_ = cross(rD_-rB_,F03_+Fin3_+G3_)+cross(rC_-rB_,-F24_)...
           +cross(rC2_-rB_,Fin2_+G2_)+Min2_;
sumM32Bz = sumM32B_(3);
F03ys=eval(solve(sumM32Bz));
F03_ = [0 F03ys 0];
% F03_ = [0,-143.783,0] (N)
```

5.2.5 Reaction Force $\mathbf{F}_{05}$

The revolute joint at E between ground 0 and the slider 5 is replaced with the joint reaction force $\mathbf{F}_{05}$ as shown in Fig. 5.6. The joint reaction force of the ground 0 on the slider 5, $\mathbf{F}_{05}$, acts at E and has two unknown components $\mathbf{F}_{05} = F_{05x}\,\mathbf{\imath} + F_{05y}\,\mathbf{J}$.

The sum of the forces on slider 5 is equal to zero, Fig. 5.6a, or

$$\sum \mathbf{F}^{(5)} = \mathbf{F}_{05} + \mathbf{F}_{in\,5} + \mathbf{G}_5 + \mathbf{F}_{45} = \mathbf{0}, \qquad (5.15)$$

Fig. 5.6 Force diagram for calculating $\mathbf{F}_{05}$

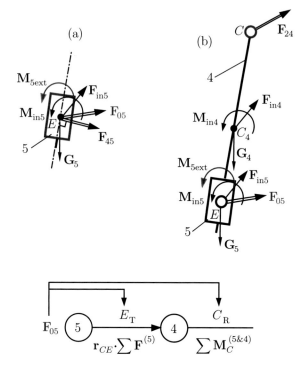

where $\mathbf{F}_{45}$ is the reaction force of the link 4 on the slider 5 and is perpendicular to the sliding direction CE, i.e., $\mathbf{F}_{45} \cdot \mathbf{r}_{CE} = 0$. If Eq. (5.15) is scalar multiplied with $\mathbf{r}_{CE}$ the following equation is obtained

$$\mathbf{r}_{CE} \cdot \sum \mathbf{F}^{(5)} = (\mathbf{F}_{05} + \mathbf{F}_{\text{in} 5} + \mathbf{G}_5) \cdot \mathbf{r}_{CE} = 0. \tag{5.16}$$

Equation (5.16) does not contain the reaction $\mathbf{F}_{45}$ because $\mathbf{F}_{45} \cdot \mathbf{r}_{CE} = 0$. For the links 5 and 4 the sum of the moments with respect to the revolute joint at C, Fig. 5.6b, gives

$$\sum \mathbf{M}_C^{(5\&4)} = \mathbf{r}_{CE} \times (\mathbf{F}_{05} + \mathbf{F}_{\text{in} 5} + \mathbf{G}_5) + \mathbf{r}_{CC_4} \times (\mathbf{F}_{\text{in} 4} + \mathbf{G}_4)$$
$$+\mathbf{M}_e + \mathbf{M}_{\text{in} 4} + \mathbf{M}_{\text{in} 5} = \mathbf{0}, \tag{5.17}$$

where the external moment on slider 5 is $\mathbf{M}_e = \mathbf{M}_{5\text{ext}}$. Equation (5.17) has only a component on $z-$axis. Equations (5.16) and (5.17) represent a system of two equations with two unknowns F_{05x} and F_{05y}. The MATLAB commands for finding $\mathbf{F}_{05}$ are:

```
F05x=sym('F05x','real');
```

```
F05y=sym('F05y','real');
F05_=[F05x, F05y, 0]; % unknown joint force
% E_P: sumF5_.CE_ = 0
% (F05_+Fin5_+G5_).CE_ = 0
SF5CE = dot(F05_+Fin5_+G5_,rE_-rC_);
% C_R: sumM54C = 0
% CE_x(F05_+Fin5_+G5_)+CC4_x(Fin4_+G4_)+Me_+Min5_+Min4_ = 0_
sumM54C_ = cross(rE_-rC_,F05_+Fin5_+G5_)+...
           cross(rC4_-rC_,Fin4_+G4_)+...
           Me_+Min5_+Min4_;
sumM54Cz = sumM54C_(3);
solF05=solve(SF5CE,sumM54Cz);
F05xs=eval(solF05.F05x);
F05ys=eval(solF05.F05y);
F05_ = [F05xs F05ys 0];
% F05_ = [-2822.194,389.528,0] (N)
```

5.2.6 Reaction Force $\mathbf{F}_{54}$ and Reaction Moment $\mathbf{M}_{54}$

The prismatic joint at E between link 4 and the slider 5 is replaced with the joint reaction force $\mathbf{F}_{54}$ and the joint reaction moment $\mathbf{M}_{54}$, as shown in Fig. 5.7.
The unknown joint reaction moment of the link 4 on the slider 5 is $\mathbf{M}_{45} = -\mathbf{M}_{54}$, has the expression $\mathbf{M}_{45} = M_{45z} \mathbf{k}$. The sum of the moments for slider 5 with respect to the revolute joint at E, Fig. 5.7a, gives

$$\sum \mathbf{M}_E^{(5)} = \mathbf{M}_{45} + \mathbf{M}_{\text{in}\,5} + \mathbf{M}_e = 0,$$

and

$$\mathbf{M}_{45} = -\mathbf{M}_{\text{in}\,5} - \mathbf{M}_e. \tag{5.18}$$

The joint reaction force of the slider 5 on link 4, $\mathbf{F}_{54}$, acts at E, has two unknown components, F_{54x}, F_{54y}, $\mathbf{F}_{54} = F_{54x} \mathbf{1} + F_{54y} \mathbf{J}$, and is perpendicular to the sliding direction CE

$$\mathbf{F}_{54} \cdot \mathbf{r}_{CE} = 0. \tag{5.19}$$

The sum of the moments on link 4 with respect to the revolute joint at C, Fig. 5.7b, gives

$$\sum \mathbf{M}_C^{(4)} = \mathbf{r}_{CE} \times \mathbf{F}_{54} + \mathbf{r}_{CC_4} \times (\mathbf{F}_{\text{in}\,4} + \mathbf{G}_4)$$
$$+\mathbf{M}_{\text{in}\,4} + \mathbf{M}_{54} = 0, \tag{5.20}$$

Fig. 5.7 Force diagram for calculating $\mathbf{F}_{45}$ and $\mathbf{M}_{45}$

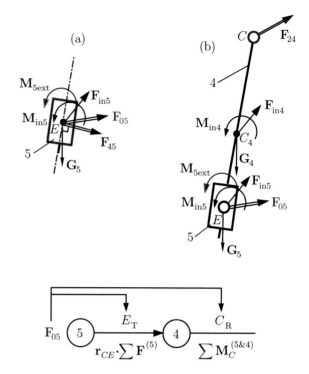

The components F_{54x} and F_{54y} are calculated from Eqs. (5.19) and (5.20). The MATLAB commands for determining $\mathbf{F}_{54}$ and $\mathbf{M}_{54}$, are:

```
% E_R: sumM5E_ = 0_
% M45_+Min5_+Me_=0_
M45_ = - (Min5_+Me_);
M54_ = -M45_;
F54x=sym('F54x','real');
F54y=sym('F54y','real');
F54_=[F54x, F54y, 0]; % unknown joint force
% F54_ perpendicular to CE_: F54_.CE_=0
PF54 = dot(F54_,rE_-rC_);
% C_R: sumM4C_ = 0_
% CE_x(F54_)+CC4_x(Fin4_+G4_)+Min4_+M54_ = 0_
sumM4C_ = cross(rE_-rC_,F54_)+...
          cross(rC4_-rC_,Fin4_+G4_)+...
          Min4_+M54_;
sumM4Cz = sumM4C_(3);
solF54=solve(PF54,sumM4Cz);
```

```
F54xs=eval(solF54.F54x);
F54ys=eval(solF54.F54y);
F54_  = [F54xs F54ys 0];
% M54_  = [0,0,999.821]  (N m)
% F54_  = [-2822.194,384.507,0]  (N)
```

5.2.7   Reaction Force $\mathbf{F}_{01}$ and Moment $\mathbf{M}_m$

The revolute joint at A between ground 0 and the driver link 1 is replaced with the joint reaction force $\mathbf{F}_{01}$ as shown in Fig. 5.8. The joint reaction force of the ground 0 on the link 1, $\mathbf{F}_{01}$, acts at A and has two unknown components $\mathbf{F}_{01} = F_{01x}\,\mathbf{i} + F_{01y}\,\mathbf{j}$. The unknown equilibrium moment $\mathbf{M}_m = M_m\,\mathbf{k}$ acts on the driver link 1. There are three unknowns and three algebraic equations must be written. The first equation is the sum of the moments on link 1 with respect to the revolute joint at B, Fig. 5.8a

$$\sum \mathbf{M}_B^{(1)} = \mathbf{r}_{BA} \times \mathbf{F}_{01} + \mathbf{r}_{BC_1} \times (\mathbf{F}_{in\,1} + \mathbf{G}_1)$$
$$+ \mathbf{M}_{in\,1} + \mathbf{M}_m = \mathbf{0}, \tag{5.21}$$

The sum of the moments on link 1 and link 2 with respect to the revolute joint at D, Fig. 5.8b, gives

$$\sum \mathbf{M}_D^{(1\&2)} = \mathbf{r}_{DA} \times \mathbf{F}_{01} + \mathbf{r}_{DC_1} \times (\mathbf{F}_{in\,1} + \mathbf{G}_1) + \mathbf{M}_{in\,1} + \mathbf{M}_m$$
$$+ \mathbf{r}_{DC} \times \mathbf{F}_{42} + \mathbf{r}_{DC_2} \times (\mathbf{F}_{in\,2} + \mathbf{G}_2) + \mathbf{M}_{in\,2} = \mathbf{0}. \tag{5.22}$$

The sum of the forces on the links 1, 2, and 3, Fig. 5.8c, gives

$$\sum \mathbf{F}^{(1\&2\&3)} = \mathbf{F}_{01} + \mathbf{F}_{in\,1} + \mathbf{G}_1 + \mathbf{F}_{42} + \mathbf{F}_{in\,2} + \mathbf{G}_2$$
$$+ \mathbf{F}_{in\,3} + \mathbf{G}_3 + \mathbf{F}_{03} = \mathbf{0}, \tag{5.23}$$

where $\mathbf{F}_{03}$ is the reaction force of the ground on slider 3 and is perpendicular to the sliding direction $x-$axis. Equation (5.23) is scalar multiplied with $\mathbf{i}$ and the following relation is obtained

$$\mathbf{i} \cdot \sum \mathbf{F}^{(1\&2\&3)} = (\mathbf{F}_{01} + \mathbf{F}_{in\,1} + \mathbf{G}_1 + \mathbf{F}_{42} + \mathbf{F}_{in\,2} + \mathbf{G}_2$$
$$+ \mathbf{F}_{in\,3} + \mathbf{G}_3) \cdot \mathbf{i} = 0. \tag{5.24}$$

From Eqs. (5.23) and (5.24) the two unknowns F_{01x} and F_{01y} are calculated. From Eq. (5.22) the unknown M_m is determined. The MATLAB commands for $\mathbf{F}_{01}$ and $\mathbf{M}_m$ are:

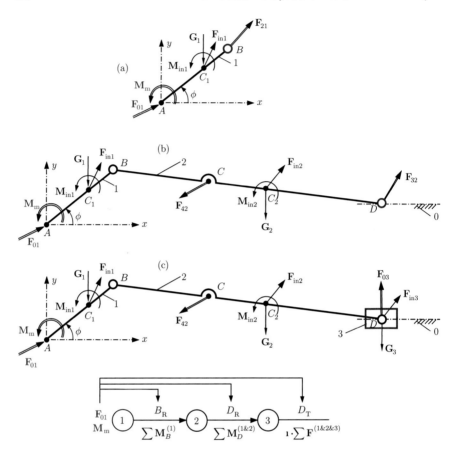

Fig. 5.8 Force diagram for calculating $\mathbf{F}_{01}$ and $\mathbf{M}_m$

```
F01x=sym('F01x','real');
F01y=sym('F01y','real');
F01_=[F01x, F01y, 0]; % unknown joint force
Mm=sym('Mm','real');
Mm_=[0, 0, Mm]; % unknown motor moment
% B_R: sumM1B_ = 0_
% BA_x(F01_)+BC1_x(Fin1_+G1_)+Min1_+Mm_ = 0_
sumM1B_ = cross(-rB_,F01_)+...
          cross(rC1_-rB_,Fin1_+G1_)+...
          Min1_+Mm_;
sumM1Bz = sumM1B_(3);
% D_R: sumM12D_ = 0_
% DA_x(F01_)+DC1_x(Fin1_+G1_)+Min1_+Mm_+
% DC_xF42_+DC2_x(Fin2_+G2_)+Min2_= 0_
```

```
sumM12D_ = cross(-rD_,F01_)+...
           cross(rC1_-rD_,Fin1_+G1_)+...
           cross(rC_-rD_,F42_)+...
           Min1_+Mm_+...
           cross(rC2_-rD_,Fin2_+G2_)+...
           Min2_;
sumM12Dz = sumM12D_(3);
% D_P: sumF123x = 0
% (F01_+Fin1_+G1_+F42_+Fin2_+G2_+Fin3_+G3_)(1)= 0
sumF123_ = (F01_+Fin1_+G1_+F42_+Fin2_+G2_+Fin3_+G3_);
sumF123x = sumF123_(1);
solF01=solve(sumM1Bz,sumM12Dz,sumF123x);
F01xs=eval(solF01.F01x);
F01ys=eval(solF01.F01y);
Ms=eval(solF01.Mm);
F01_ = [F01xs F01ys 0];
Mm_ = [0 0 Ms];
% F01_ = [1962.873,-571.442,0] (N)
% Mm_ = [0,0,-358.628] (N m)
```

The results are identical with the results obtained with the classical method.

5.3 Problems

5.1 Find the positions, velocities, accelerations, and joint forces for the mechanism
described in Problem 4.1 using the contour methods. Compare with the results
obtained with the classic method.

5.2 Repeat Problem 4.2 using the contour methods. Compare with the results
obtained with the classic method.

5.3 Find the kinematics and the dynamics of the mechanism in Problem 4.3 using
the contour methods. Compare with the results obtained with the classic method.

5.4 Repeat Problem 4.4 using the contour methods. Compare with the results
obtained with the classic method.

5.5 Repeat Problem 4.5 using the contour methods. Compare with the results
obtained with the classic method.

References

1. E.A. Avallone, T. Baumeister, A. Sadegh, *Marks' Standard Handbook for Mechanical Engineers*, 11th edn. (McGraw-Hill Education, New York, 2007)
2. I.I. Artobolevski, *Mechanisms in Modern Engineering Design* (MIR, Moscow, 1977)

3. M. Atanasiu, *Mechanics (Mecanica)* (EDP, Bucharest, 1973)
4. H. Baruh, *Analytical Dynamics* (WCB/McGraw-Hill, Boston, 1999)
5. A. Bedford, W. Fowler, *Dynamics* (Addison Wesley, Menlo Park, CA, 1999)
6. A. Bedford, W. Fowler, *Statics* (Addison Wesley, Menlo Park, CA, 1999)
7. F.P. Beer et al., *Vector Mechanics for Engineers: Statics and Dynamics* (McGraw-Hill, New York, NY, 2016)
8. M. Buculei, D. Bagnaru, G. Nanu, D.B. Marghitu, *Computing Methods in the Analysis of the Mechanisms with Bars* (Scrisul Romanesc, Craiova, 1986)
9. M.I. Buculei, *Mechanisms* (University of Craiova Press, Craiova, Romania, 1976)
10. D. Bolcu, S. Rizescu, *Mecanica* (EDP, Bucharest, 2001)
11. J. Billingsley, *Essential of Dynamics and Vibration* (Springer, 2018)
12. R. Budynas, K.J. Nisbett, *Shigley's Mechanical Engineering Design*, 9th edn. (McGraw-Hill, New York, 2013)
13. J..A. Collins, H.R. Busby, G.H. Staab, *Mechanical Design of Machine Elements and Machines*, 2nd edn. (Wiley, 2009)
14. M. Crespo da Silva, *Intermediate Dynamics for Engineers* (McGraw-Hill, New York, 2004)
15. A.G. Erdman, G.N. Sandor, *Mechanisms Design* (Prentice-Hall, Upper Saddle River, NJ, 1984)
16. M. Dupac, D.B. Marghitu, *Engineering Applications: Analytical and Numerical Calculation with MATLAB* (Wiley, Hoboken, NJ, 2021)
17. A. Ertas, J.C. Jones, *The Engineering Design Process* (Wiley, New York, 1996)
18. F. Freudenstein, An application of Boolean algebra to the motion of epicyclic drivers. J. Eng. Ind. pp. 176–182 (1971)
19. J.H. Ginsberg, *Advanced Engineering Dynamics* (Cambridge University Press, Cambridge, 1995)
20. H. Goldstein, *Classical Mechanics* (Addison-Wesley, Redwood City, CA, 1989)
21. D.T. Greenwood, *Principles of Dynamics* (Prentice-Hall, Englewood Cliffs, NJ, 1998)
22. A.S. Hall, A.R. Holowenko, H.G. Laughlin, *Schaum's Outline of Machine Design* (McGraw-Hill, New York, 2013)
23. B.G. Hamrock, B. Jacobson, S.R. Schmid, *Fundamentals of Machine Elements* (McGraw-Hill, New York, 1999)
24. R.C. Hibbeler, *Engineering Mechanics: Dynamics* (Prentice Hall, 2010)
25. T.E. Honein, O.M. O'Reilly, On the Gibbs-Appell equations for the dynamics of rigid bodies. J. Appl. Mech. **88**, 074501–1 (2021)
26. R.C. Juvinall, K.M. Marshek, *Fundamentals of Machine Component Design*, 5th edn. (Wiley, New York, 2010)
27. T.R. Kane, *Analytical Elements of Mechanics*, vol. 1 (Academic Press, New York, 1959)
28. T.R. Kane, *Analytical Elements of Mechanics*, vol. 2 (Academic Press, New York, 1961)
29. T.R. Kane, D.A. Levinson, The use of Kane's dynamical equations in robotics. MIT Int. J. Robot. Res. **3**, 3–21 (1983)
30. T.R. Kane, P.W. Likins, D.A. Levinson, *Spacecraft Dynamics* (McGraw-Hill, New York, 1983)
31. T.R. Kane, D.A. Levinson, *Dynamics* (McGraw-Hill, New York, 1985)
32. K. Lingaiah, *Machine Design Databook*, 2nd edn. (McGraw-Hill Education, New York, 2003)
33. N.I. Manolescu, F. Kovacs, A. Oranescu, *The Theory of Mechanisms and Machines* (EDP, Bucharest, 1972)
34. D.B. Marghitu, *Mechanical Engineer's Handbook* (Academic Press, San Diego, CA, 2001)
35. D.B. Marghitu, M.J. Crocker, *Analytical Elements of Mechanisms* (Cambridge University Press, Cambridge, 2001)
36. D.B. Marghitu, *Kinematic Chains and Machine Component Design* (Elsevier, Amsterdam, 2005)
37. D.B. Marghitu, *Mechanisms and Robots Analysis with MATLAB* (Springer, New York, N.Y., 2009)
38. D.B. Marghitu, M. Dupac, *Advanced Dynamics: Analytical and Numerical Calculations with MATLAB* (Springer, New York, N.Y., 2012)
39. D.B. Marghitu, M. Dupac, H.M. Nels, *Statics with MATLAB* (Springer, New York, N.Y., 2013)

40. D.B. Marghitu, D. Cojocaru, *Advances in Robot Design and Intelligent Control* (Springer International Publishing, Cham, Switzerland, 2016), pp. 317–325

41. D.J. McGill, W.W. King, *Engineering Mechanics: Statics and an Introduction to Dynamics* (PWS Publishing Company, Boston, 1995)

42. J.L. Meriam, L.G. Kraige, *Engineering Mechanics: Dynamics* (Wiley, New York, 2007)

43. C.R. Mischke, Prediction of stochastic endurance strength. Trans. ASME, J. Vib. Acoust. Stress Reliab. Des. **109**(1), 113–122 (1987)

44. L. Meirovitch, *Methods of Analytical Dynamics* (Dover, 2003)

45. R.L. Mott, *Machine Elements in Mechanical Design* (Prentice Hall, Upper Saddle River, NJ, 1999)

46. W.A. Nash, *Strength of Materials*. Schaum's Outline Series (McGraw-Hill, New York, 1972)

47. R.L. Norton, *Machine Design* (Prentice-Hall, Upper Saddle River, NJ, 1996)

48. R.L. Norton, *Design of Machinery* (McGraw-Hill, New York, 1999)

49. O.M. O'Reilly, *Intermediate Dynamics for Engineers Newton-Euler and Lagrangian Mechanics* (Cambridge University Press, UK, 2020)

50. O.M. O'Reilly, *Engineering Dynamics: A Primer* (Springer, NY, 2010)

51. W.C. Orthwein, *Machine Component Design* (West Publishing Company, St. Paul, 1990)

52. L.A. Pars, *A Treatise on Analytical Dynamics* (Wiley, New York, 1965)

53. F. Reuleaux, *The Kinematics of Machinery* (Dover, New York, 1963)

54. D. Planchard, M. Planchard, *SolidWorks 2013 Tutorial with Video Instruction* (SDC Publications, 2013)

55. I. Popescu, *Mechanisms* (University of Craiova Press, Craiova, Romania, 1990)

56. I. Popescu, C. Ungureanu, *Structural Synthesis and Kinematics of Mechanisms with Bars* (Universitaria Press, Craiova, Romania, 2000)

57. I. Popescu, L. Luca, M. Cherciu, D.B. Marghitu, *Mechanisms for Generating Mathematical Curves* (Springer Nature, Switzerland, 2020)

58. C.A. Rubin, *The Student Edition of Working Model* (Addison-Wesley Publishing Company, Reading, MA, 1995)

59. J. Ragan, D.B. Marghitu, Impact of a kinematic link with MATLAB and solidworks. Appl. Mech. Mater. **430**, 170–177 (2013)

60. J. Ragan, D.B. Marghitu, MATLAB dynamics of a free link with elastic impact, in *International Conference on Mechanical Engineering, ICOME 2013*, May 16–17, 2013, Craiova, Romania (2013)

61. J.C. Samin, P. Fisette, *Symbolic Modeling of Multibody Systems* (Kluwer, 2003)

62. A.A. Shabana, *Computational Dynamics* (Wiley, New York, 2010)

63. I.H. Shames, *Engineering Mechanics - Statics and Dynamics* (Prentice-Hall, Upper Saddle River, NJ, 1997)

64. J.E. Shigley, C.R. Mischke, *Mechanical Engineering Design* (McGraw-Hill, New York, 1989)

65. J.E. Shigley, C.R. Mischke, R.G. Budynas, *Mechanical Engineering Design*, 7th edn. (McGraw-Hill, New York, 2004)

66. J.E. Shigley, J.J. Uicker, *Theory of Machines and Mechanisms* (McGraw-Hill, New York, 1995)

67. D. Smith, *Engineering Computation with MATLAB* (Pearson Education, Upper Saddle River, NJ, 2008)

68. R.W. Soutas-Little, D.J. Inman, *Engineering Mechanics: Statics and Dynamics* (Prentice-Hall, Upper Saddle River, NJ, 1999)

69. J. Sticklen, M.T. Eskil, *An Introduction to Technical Problem Solving with MATLAB* (Great Lakes Press, Wildwood, MO, 2006)

70. A.C. Ugural, *Mechanical Design* (McGraw-Hill, New York, 2004)

71. R. Voinea, D. Voiculescu, V. Ceausu, *Mechanics (Mecanica)* (EDP, Bucharest, 1983)

72. K.J. Waldron, G.L. Kinzel, *Kinematics, Dynamics, and Design of Machinery* (Wiley, New York, 1999)

73. J.H. Williams Jr., *Fundamentals of Applied Dynamics* (Wiley, New York, 1996)

74. C.E. Wilson, J.P. Sadler, *Kinematics and Dynamics of Machinery* (Harper Collins College Publishers, New York, 1991)

75. H.B. Wilson, L.H. Turcotte, D. Halpern, *Advanced Mathematics and Mechanics Applications Using MATLAB* (Chapman & Hall/CRC, 2003)
76. S. Wolfram, *Mathematica* (Wolfram Media/Cambridge University Press, Cambridge, 1999)
77. National Council of Examiners for Engineering and Surveying (NCEES), *Fundamentals of Engineering. Supplied-Reference Handbook*, Clemson, SC (2001)
78. * * * , *The Theory of Mechanisms and Machines* (Teoria mehanizmov i masin), Vassaia scola, Minsk, Russia (1970)
79. eCourses - University of Oklahoma: http://ecourses.ou.edu/home.htm
80. https://www.mathworks.com
81. http://www.eng.auburn.edu/~marghitu/
82. https://www.solidworks.com
83. https://www.wolfram.com

Chapter 6
Dyad Routines for Mechanisms

Abstract MATLAB functions are developed for position, velocity, and acceleration of dyads and kinematic links. The dynamics of mechanisms with one and two dyads are analyzed. For force analysis D'Alembert principle is employed.

With the help of dyads or system groups a mechanism can be assembled. MAT-LAB functions will be developed for dyads and driver links for positions, velocities, accelerations, and joint forces.

6.1 Driver Link

A rotating rigid body (i) is shown in Fig. 6.1a with two points A_i and P on the link (i), $A_i \in (i)$, $P \in (i)$. The input data for the function pvaR are the position, velocity, and acceleration of point A_i, an angular position, angular velocity, and angular acceleration of link (i):

```
function out = pvaR(rAi_, vAi_, aAi_, li, phi, omega, alpha)
% rAi_ is the position vector of Ai
% vAi_ is the velocity vector of Ai
% aAi_ is the acceleration vector of Ai
% li is the length from Ai to P: li=AiP
% phi is the angle of link (i) with x-axis
% omega is the angular velocity of  link (i),  AiP
% alpha is the angular acceleration of link (i),  AiP
```

The position, velocity, and acceleration of point P are calculated with:

```
omega_ = [0 0 omega]; % angular velocity vector of AiP
alpha_ = [0 0 alpha]; % angular acceleration vector of AiP
% position of P
xP = rAi_(1)+li*cos(phi);
yP = rAi_(2)+li*sin(phi);
```

© The Author(s), under exclusive license to Springer Nature Switzerland AG 2022
D. B. Marghitu et al., *Mechanical Simulation with MATLAB®*,
Springer Tracts in Mechanical Engineering,
https://doi.org/10.1007/978-3-030-88102-3_6

```
rP_ = [xP, yP, 0]; % position vector of P
% velocity of P
vP_ = vAi_ + cross(omega_, rP_-rAi_);
vPx = vP_(1); % x-component of velocity of P
vPy = vP_(2); % y-component of velocity of P
% acceleration of P
aP_ = aAi_ + cross(alpha_, rP_-rAi_)-omega^2*(rP_-rAi_);
aPx = aP_(1); % x-component of acceleration of P
aPy = aP_(2); % y-component of acceleration of P
out = [xP, yP, vPx, vPy, aPx, aPy];
end
```

The components of the position, velocity, and acceleration of point P are the outputs of the function pvaR.

6.2 Position Analysis

Figure 6.1b shows three points P, Q, and R on the same line. The inputs for the function pos3P are the position vectors of points P, Q and the length from P to R. If R is on the same direction from P to Q then a positive PR is selected, Fig. 6.1b. If R is opposite to the direction from P to Q then a negative PR is selected:

```
function out = pos3P(rP_, rQ_ , PR)
% rP_ is position vector of P
% rQ_ is position vector of Q
% PR  is length of the segment PR
% select
% +PR if R is on the same direction from P to Q  P->R->Q or P->Q->R
% -PR if R is opposite to the direction from P to Q  R->P->Q

% coordinates of P
xP = rP_(1);
yP = rP_(2);
% coordinates of Q
xQ = rQ_(1);
yQ = rQ_(2);
```

The coordinates of point R, xR and yR are calculated as

```
% orientation of line PQ
PQ = norm([xQ-xP, yQ-yP]);
cPQ = (xQ-xP)/PQ;
sPQ = (yQ-yP)/PQ;
% coordinates of R
xR = xP + PR*cPQ;
```

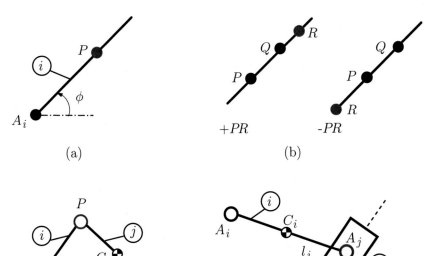

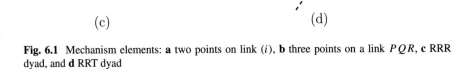

Fig. 6.1 Mechanism elements: **a** two points on link (i), **b** three points on a link PQR, **c** RRR dyad, and **d** RRT dyad

```
yR = yP + PR*sPQ;
out = [xR yR];
```

Position RRR Dyad

Figure 6.1c shows a RRR dyad. The inputs are the position vectors of A_i and A_j and the lengths of the links (i) and (j):

```
function out = posRRR(rAi_, rAj_, li, lj)
% rAi_ is the position vector of Ai: rAi_ = [xAi yAi 0]
% rAj_ is the position vector of Aj: rAj_ = [xAj yAj 0]
% li is the length of link (i): li=AiP
% lj is the length of link (j): lj=AjP
% components of position of Ai
xAi = rAi_(1);
yAi = rAi_(2);
% components of position of Aj
```

```
xAj = rAj_(1);
yAj = rAj_(2)
```

The outputs are the positions of the point P and are calculated with:

```
syms xP yP
eqRRRi = (xAi-xP)^2+(yAi-yP)^2-li^2;
eqRRRj = (xAj-xP)^2+(yAj-yP)^2-lj^2;
solRRR = solve(eqRRRi, eqRRRj);
xPpos = eval(solRRR.xP);
yPpos = eval(solRRR.yP);
xP1 = xPpos(1); xP2 = xPpos(2);
yP1 = yPpos(1); yP2 = yPpos(2);

out = [xP1, yP1, xP2, yP2];
% rPI_  = [xP1 yP1 0] solution I  for P
% rPII_ = [xP2 yP2 0] solution II for P
```

The routine posRRR calculates two solutions for the joint P: xP1, yP1 and xP2, yP2.

Position RRT Dyad

For the RRT dyad shown in Fig. 6.1d the inputs are the position vectors of A_i and A_s, the length of the link (i), and the orientation θ of the slider:

```
function out = posRRT(rAi_, rAs_, li, Theta)
% rAi_ is the position vector of Ai
% rAs_ is the position vector of a point As on the sliding direction
% li is the length of link (i) li=AiAj
% Theta is the angle of the sliding direction with x-axis

% components of position of Ai
xAi = rAi_(1);
yAi = rAi_(2);
% components of position of As
xAs = rAs_(1);
yAs = rAs_(2);

% rAj_ = [xAj yAj 0] position vector of Aj=Cj on link (j)
```

There are two solutions for the components of the joint A_j:

```
% syms xAj yAj
if Theta==pi/2 || Theta==3*pi/2
   xAj1 = xAs;
   xAj2 = xAs;
   xAj = xAs;
```

```
    syms yAj
    eqRRT = (xAi-xAj)^2+(yAi-yAj)^2-li^2;
    solRRT = solve(eqRRT);
    yAj1 = eval(solRRT(1));
    yAj2 = eval(solRRT(2));

elseif Theta==0 || Theta==pi
    yAj1 = yAs; yAj2 = yAs;
    yAj = yAs;
    syms xAj
    eqRRT = (xAi-xAj)^2+(yAi-yAj)^2-li^2;
    solRRT = solve(eqRRT);
    xAj1 = eval(solRRT(1));
    xAj2 = eval(solRRT(2));
else
    syms xAj yAj
    eqRRT1 = (xAi-xAj)^2 + (yAi-yAj)^2 - li^2;
    eqRRT2 = tan(Theta) - (yAj-yAs)/(xAj-xAs);
    solRRT = solve(eqRRT1, eqRRT2);
    xAjpos = eval(solRRT.xAj);
    yAjpos = eval(solRRT.yAj);
    xAj1 = xAjpos(1); xAj2 = xAjpos(2);
    yAj1 = yAjpos(1); yAj2 = yAjpos(2);
end
out = [xAj1 yAj1 xAj2 yAj2];
% rAjI_  = [xAj1 yAj1 0] solution I for Aj
% rAjII_ = [xAj2 yAj2 0] solution II for Aj
```

6.3 Velocity Analysis

Linear Velocity and Acceleration

The routine linvelacc.m calculates the velocity and acceleration of the point Q on a link PQ, when the velocity and acceleration of the point P are given, Fig. 6.1b. The input data are:

```
function out=...
 linvelacc(rP_,rQ_,vP_,aP_,omega,alpha)
% rP_ is the position vector of P
% rQ_ is the position vector of Q
% vP_ is the velocity vector of P
% aP_ is the acceleration vector of P
% omega is the angular velocity of segment PQ
```

```
% alpha is the angular acceleration of segment PQ
```

The components of the velocity and acceleration of point Q are calculated with:

```
% position, velocity, acceleration of P
xP = rP_(1);
yP = rP_(2);
vPx = vP_(1);
vPy = vP_(2);
aPx = aP_(1);
aPy = aP_(2);

% position of Q
xQ = rQ_(1);
yQ = rQ_(2);

% velocity of Q
vQx = vPx - omega*(yQ - yP);
vQy = vPy + omega*(xQ - xP);

% acceleration of Q
aQx = aPx - alpha*(yQ - yP) - omega^2*(xQ - xP);
aQy = aPy + alpha*(xQ - xP) - omega^2*(yQ - yP);

% components of linear velocity and acceleration of Q
out = [vQx vQy aQx aQy];
```

Angular Velocity and Acceleration

The routine angvelacc.m calculates the angular velocity and acceleration of a segment PQ when the linear velocity and acceleration of P and Q are given, Fig. 6.1b:

```
function out=angvelacc(rP_, rQ_, vP_, vQ_, aP_, aQ_)
% P and Q are the input points
% rP_ is the position vector of P
% rQ_ is the position vector of Q

% vP_ is the velocity vector of P
% vQ_ is the velocity vector of Q

% aP_ is the acceleration vector of P
% aQ_ is the acceleration vector of Q
```

The angular velocity and acceleration of segment PQ are calculated with:

```
% coordinates position, velocity, acceleration of P
xP = rP_(1);
yP = rP_(2);
vPx = vP_(1);
vPy = vP_(2);
aPx = aP_(1);
aPy = aP_(2);

% coordinates position, velocity, acceleration of Q
xQ = rQ_(1);
yQ = rQ_(2);
vQx = vQ_(1);
vQy = vQ_(2);
aQx = aQ_(1);
aQy = aQ_(2);

% angle between PQ and x-axis
theta = atan((yP-yQ)/(xP-xQ));

% angular velocity of PQ
dtheta = ...
  (cos(theta)*(vPy-vQy)-sin(theta)*(vPx-vQx))/...
  (sin(theta)*(yP-yQ)+cos(theta)*(xP-xQ));

% angular acceleration of PQ
ddtheta = ...
  (cos(theta)*(aPy-aQy)-sin(theta)*(aPx-aQx)...
  -(cos(theta)*dtheta*(yP-yQ)+2*sin(theta)*(vPy-vQy)...
  -sin(theta)*dtheta*(xP-xQ)+2*cos(theta)*(vPx-vQx))*dtheta/...
  (cos(theta)*(xP-xQ)+sin(theta)*(yP-yQ));

out = [dtheta ddtheta];
```

RRR dyad
The routine `velaccRRR.m` calculates the velocity and acceleration of RRR dyad, Fig. 6.1c. The input data are:

```
function out=velaccRRR...
(rAi_,rAj_,rP_,vAi_,vAj_,aAi_,aAj_)
% rAi_ is the position vector of Ai
% rAj_ is the position vector of Aj
% rP_  is the position vector of P
% vAi_ is the velocity vector of Ai
```

```
% vAj_ is the velocity vector of Aj
% aAi_ is the acceleration vector of Ai
% aAj_ is the acceleration vector of Aj

% components of position, velocity, acceleration of Ai
xAi = rAi_(1);
yAi = rAi_(2);
vAix = vAi_(1);
vAiy = vAi_(2);
aAix = aAi_(1);
aAiy = aAi_(2);

% components of position, velocity, acceleration of Aj
xAj = rAj_(1);
yAj = rAj_(2);
vAjx = vAj_(1);
vAjy = vAj_(2);
aAjx = aAj_(1);
aAjy = aAj_(2);

% components of position vector of P
xP = rP_(1);
yP = rP_(2);
```

The velocity and the acceleration components of the joint *P* are:

```
% velocity of P
syms vPxs vPys
eqRRR1v =...
(xAi-xP)*(vAix-vPxs)+(yAi-yP)*(vAiy-vPys);
eqRRR2v =...
(xAj-xP)*(vAjx-vPxs)+(yAj-yP)*(vAjy-vPys);

solRRRv=solve(eqRRR1v, eqRRR2v, vPxs, vPys);
vPx = eval(solRRRv.vPxs);
vPy = eval(solRRRv.vPys);

% acceleration of P
syms aPxs aPys
eqRRR1a =...
(xAi-xP)*(aAix-aPxs)+(vAix-vPx)^2+...
(yAi-yP)*(aAiy-aPys)+(vAiy-vPy)^2;
eqRRR2a =...
(xAj-xP)*(aAjx-aPxs)+(vAjx-vPx)^2+...
(yAj-yP)*(aAjy-aPys)+(vAjy-vPy)^2;
```

```
solRRRa=solve(eqRRR1a, eqRRR2a, aPxs, aPys);
aPx = eval(solRRRa.aPxs);
aPy = eval(solRRRa.aPys);

% components of velocity and acceleration of P
out = [vPx vPy aPx aPy];
```

RRT dyad

For the RRT dyad, Fig. 6.1d, the routine `velaccRRT.m` is employed with the following input data:

```
function out=velaccRRT(rAi_,rAs_,rAj_,...
 vAi_,vAs_,aAi_,aAs_,dtheta,ddtheta)

% rAi_ is the position vector of Ai
% rAs_ is the position vector of As on the sliding direction
% rAj_ is the position vector of Aj=Cj
% vAi_ is the velocity vector of Ai
% vAs_ is the velocity vector of As on the sliding direction
% aAi_ is the acceleration vector of Ai
% aAs_ is the acceleration vector of As on the sliding direction
% dtheta is the angular velocity of the sliding direction
% ddtheta is the angular acceleration of the sliding direction

% components of position, velocity, acceleration of Ai
xAi = rAi_(1);
yAi = rAi_(2);
vAix = vAi_(1);
vAiy = vAi_(2);
aAix = aAi_(1);
aAiy = aAi_(2);

% components of position, velocity, acceleration of As
% position, velocity, acceleration of As
% As is a point on the sliding direction
xAs = rAs_(1);
yAs = rAs_(2);
vAsx = vAs_(1);
vAsy = vAs_(2);
aAsx = aAs_(1);
aAsy = aAs_(2);

% position of Aj
xAj = rAj_(1);
yAj = rAj_(2);

theta = atan((yAj-yAs)/(xAj-xAs));
```

The outputs are the components of velocity and acceleration of A_j:

```
% calculate velocity of Aj
syms vAjxSol vAjySol
eqRRT1v =...
(xAi-xAj)*(vAix-vAjxSol)+(yAi-yAj)*(vAiy-vAjySol);
eqRRT2v =...
sin(theta)*(vAjxSol-vAsx)+cos(theta)*dtheta*(xAj-xAs)...
-cos(theta)*(vAjySol-vAsy)+sin(theta)*dtheta*(yAj-yAs)';
solRRTv = solve(eqRRT1v, eqRRT2v);
vAjx = eval(solRRTv.vAjxSol);
vAjy = eval(solRRTv.vAjySol);

% calculate acceleration of Aj
syms aAjxSol aAjySol
eqRRT1a =...
(xAi-xAj)*(aAix-aAjxSol)+(vAix-vAjx)^2+...
(yAi-yAj)*(aAiy-aAjySol)+(vAiy-vAjy)^2;
eqRRT2a =...
sin(theta)*(aAjxSol-aAsx)-cos(theta)*(aAjySol-aAsy)...
+(2*cos(theta)*(vAjx-vAsx)-sin(theta)*dtheta*(xAj-xAs)...
+2*sin(theta)*(vAjy-vAsy)...
+cos(theta)*dtheta*(yAj-yAs))*dtheta...
+(cos(theta)*(xAj-xAs)+sin(theta)*(yAj-yAs))*ddtheta;
solRRTa = solve(eqRRT1a, eqRRT2a);
aAjx = eval(solRRTa.aAjxSol);
aAjy = eval(solRRTa.aAjySol);

% components of velocity and acceleration of Aj
out = [vAjx vAjy aAjx aAjy];
```

6.4   Force Analysis

RRR dyad
The routine dforceRRR.m is used to calculate the joint reaction forces for RRR
dyad shown in Fig. 6.2a. The input data are:

```
% dforceRRR.m
% dynamic forces analysis
% RRR dyad: Ai->R   P->R   Aj->R
% link (i) with CM at Ci
% link (j) with CM at Cj
function out=dforceRRR...
```

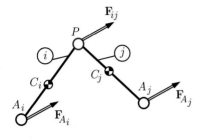

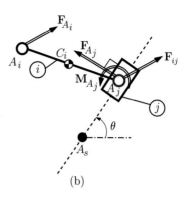

(a)

(b)

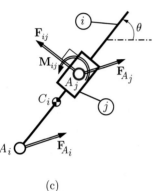

(c)

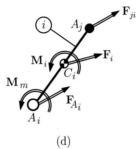

(d)

Fig. 6.2 Joint reaction forces for: **a** RRR, **b** RRT, **c** RTR, **d** kinematic element

```
(rAi_,rAj_,rP_,rCi_,rCj_,Fi_,Mi_,Fj_,Mj_)
% rAi_ is the position vector of Ai
% rAj_ is the position vector of Aj
% rP_  is the position vector of P
% rCi_ is the position vector of CM of (i)
% rCj_ is the position vector of CM of (j)
% Fi_ is the total force on (i) at Ci including
%      the inertia force at Ci and
%      the gravitational force at Ci
% Mi_ is the total moment on (i) including
%      the moment of inertia on (i)
% Fj_ is the total force on (j) at Cj including
%      the inertia force at Cj and
%      the gravitational force at Cj
% Mj_ is the total moment on (j) including
%      the moment of inertia on (j)
```

The unknowns are:

```
syms FAix FAiy FAjx FAjy
% FAix is the reaction force on (i) at Ai on x-axis
% FAiy is the reaction force on (i) at Ai on y-axis
FAi_=[FAix, FAiy, 0]; % unknown joint force  at Ai
% FAjx is the reaction force on (j) at Aj on x-axis
% FAjy is the reaction force on (j) at Aj on y-axis
FAj_=[FAjx, FAjy, 0]; % unknown joint force  at Aj
```

The joint reaction forces are calculated using D'Alembert principle:

```
% sum forces of (i) and (j)
% FAi_ + Fi_ + FAj_ + Fj_ = 0_
eqF_ = FAi_+Fi_+FAj_+Fj_;
eqFx = eqF_(1); % (1)
eqFy = eqF_(2); % (2)

rPCi_ = rCi_ - rP_; % position vector PCi_
rPCj_ = rCj_ - rP_; % position vector PCj_
rPAi_ = rAi_ - rP_; % position vector PAi_
rPAj_ = rAj_ - rP_; % position vector PAj_

% sum moments of (i) about P
% PCi_ x Fi_ + PAi_ x FAi_ + Mi_ = 0_
eqMi_ = cross(rPCi_,Fi_)+cross(rPAi_,FAi_)+Mi_;
eqMiz = eqMi_(3); % (3)

% sum moments of (j) about P
% PCj_ x Fj_ + PAj_ x FAj_ + Mj_ = 0_
eqMj_ = cross(rPCj_,Fj_)+cross(rPAj_,FAj_)+Mj_;
eqMjz = eqMj_(3); % (4)

% Eqs.(1)-(4) =>
sol=solve(eqFx, eqFy, eqMiz, eqMjz);
FAixs=eval(sol.FAix);
FAiys=eval(sol.FAiy);
FAjxs=eval(sol.FAjx);
FAjys=eval(sol.FAjy);

FAis_ = [FAixs, FAiys, 0];
FAjs_ = [FAjxs, FAjys, 0];

% sum forces of (i)
% FAi_ + Fi_ + Fji_ = 0_ =>
```

```
% Fji_ is the reaction force of (j) on (i) at P
Fji_ = - Fi_ - FAis_;
Fjix = Fji_(1); % x-component of Fji_
Fjiy = Fji_(2); % y-component of Fji_

out = [FAixs,FAiys,FAjxs,FAjys,Fjix,Fjiy];
```

RRT dyad

For the RRT dyad shown in Fig. 6.2b the input data are:

```
% dforceRRT.m
% dynamic forces analysis
% RRT dyad: Ai->R  Aj->R Aj->T
% slider (j) with CM at Aj=Cj
function out=dforceRRT...
(rAi_,rAs_,rAj_,rCi_,Fi_,Mi_,Fj_,Mj_)
% rAi_ is the position vector of Ai
% rAs_ is the position vector of As on the sliding direction
% rAj_ is the position vector of Aj=Cj
% rCi_ is the position vector of Ci
% Fi_ is the total force on (i) at Ci including
%      the inertia force at Ci and
%      the gravitational force at Ci
% Mi_ is the total moment on (i) including
%      the moment of inertia on (i)
% Fj_ is the total force on (j) at Aj=Cj including
%      the inertia force at Cj and
%      the gravitational force at Cj
% Mj_ is the total moment on (j) including
%      the moment of inertia on (j)
```

The joint reaction forces and moments are calculated with:

```
syms FAix FAiy FAjx FAjy
% FAix rection force on (i) at Ai on x-axis
% FAiy rection force on (i) at Ai on y-axis
FAi_=[FAix, FAiy, 0]; % unknown joint force  at Ai
% FAjx rection force on (j) at Aj on x-axis
% FAjy rection force on (j) at Aj on y-axis
FAj_=[FAjx, FAjy, 0]; % unknown joint force  at Aj

% sum forces on (i) and (j)
% FAi_ + Fi_ + FAj_ + Fj_ = 0_
eqF_ = FAi_+Fi_+FAj_+Fj_;
eqFx = eqF_(1); % (1)
```

```
eqFy = eqF_(2); % (2)


rAjCi_ = rCi_ - rAj_; % position vector AjCi_
rAjAi_ = rAi_ - rAj_; % position vector AjAi_
rAjAs_ = rAs_ - rAj_; % position vector AjAs_

% FAj_ projection onto rAjAs_ is zero: FAj_.AjAs_ = 0
eqFAjAs = FAj_*rAjAs_.'; % (3)

% sum of moments of (i) about Aj
% AjCi_ x Fi_ + AjAi_ x FAi_ + Mi_ = 0_
eqMi_ = cross(rAjCi_,Fi_)+cross(rAjAi_,FAi_)+Mi_;
eqMiz = eqMi_(3); % (4)

% Eqs.(1)-(4) =>
sol=solve(eqFx, eqFy, eqMiz, eqFAjAs);
FAixs=eval(sol.FAix);
FAiys=eval(sol.FAiy);
FAjxs=eval(sol.FAjx);
FAjys=eval(sol.FAjy);

% sum of moments of (j) about Aj
% MAj_ + Mj_ = 0_
% MAj_ is the reaction moment on (j) of the sliding direction
MAjz = - Mj_(3);

% numerical value
FAjs_=[FAjxs, FAjys, 0];

% sum forces of (j)
% FAj_ + Fj_ + Fij_ = 0_ =>
% Fij_ is the reaction force of (i) on (j) at Aj
Fij_ = - Fj_ - FAjs_;
Fijx = Fij_(1);
Fijy = Fij_(2);

out = [FAixs,FAiys,FAjxs,FAjys,Fijx,Fijy,MAjz];
```

RTR dyad

The force analysis for the RTR dyad, shown in Fig. 6.2c, is described by the MATLAB commands:

```
% dforceRTR.m
% forces RTR dyad: Ai->R  Aj->T Aj->R
% slider (j) with CM at Aj=Cj
```

```
function out=dforceRTR...
(rAi_,rAj_,rCi_,Fi_,Mi_,Fj_,Mj_)
% rAi_ is the position vector of Ai
% rAj_ is the position vector of Aj=Cj
% rCi_ is the position vector of Ci
% Fi_ is the total force on (i) at Ci including
%      the inertia force at Ci and
%      the gravitational force at Ci
% Mi_ is the total moment on (i) including
%      the moment of inertia on (i)
% Fj_ is the total force on (j) at Aj=Cj including
%      the inertia force at Cj and
%      the gravitational force at Cj
% Mj_ is the total moment on (j) including
%      the moment of inertia on (j)

syms FAix FAiy FAjx FAjy
% FAix is the reaction force on (i) at Ai on x-axis
% FAiy is the reaction force on (i) at Ai on y-axis
FAi_=[FAix, FAiy, 0]; % unknown joint force  at Ai
% FAjx is the reaction force on (j) at Aj on x-axis
% FAjy is the reaction force on (j) at Aj on y-axis
FAj_=[FAjx, FAjy, 0]; % unknown joint force  at Aj

% sum forces on (i) and (j)
% FAi_ + Fi_ + FAj_ + Fj_ = 0_
eqF_ = FAi_+Fi_+FAj_+Fj_;
eqFx = eqF_(1); % (1)
eqFy = eqF_(2); % (2)


rAiAj_ = rAj_ - rAi_; % vector AiAj_
% sum of forces of (j) projection onto rAiAj_
% (FAj_+Fj_+ Fij_).AiAj_ = 0 and Fij_.AiAj_ = 0
% Fij reaction force of (i) on (j) at Aj
% Fij perpendicular on AiAj_
eqFi = (FAj_+Fj_)*rAiAj_.'; % (3)

rAiCi_ = rCi_ - rAi_; % vector AiCi_
% sum of moments of (i) and (j) about Ai
% AiCi_ x Fi_ + AiAj_ x (Fj_+FAj_) + Mi_ + Mj_
eqMij_ = cross(rAiCi_,Fi_)+cross(rAiAj_,Fj_+FAj_)+Mi_+Mj_;
eqMijz = eqMij_(3); % (4)

% Eqs(1)(2)(3)(4) =>
```

```
sol=solve(eqFx, eqFy, eqFi, eqMijz);
FAixs=eval(sol.FAix);
FAiys=eval(sol.FAiy);
FAjxs=eval(sol.FAjx);
FAjys=eval(sol.FAjy);

FAjs_=[FAjxs, FAjys, 0];

% sum of forces of (j)
% FAj_ + Fj_ + Fij_ = 0_ ;
% Fij_ = -Fji_ reaction force of (i) on (j) at Aj
Fij_ = - Fj_ - FAjs_;
Fijx = Fij_(1);
Fijy = Fij_(2);

% sum of moments of (j) about Aj
% Mij_ + Mj_= 0_; =>
% Mij reaction moment of (i) on (j)
Mijz = - Mj_(3);

out = [FAixs,FAiys,FAjxs,FAjys,Fijx,Fijy,Mijz];
end
```

For the kinematic link shown in Fig. 6.2d, the MATLAB commands for dynamic force analysis are:

```
% dforceR.m
% forces on link (i)
function out=dforceR(rAi_,rAj_,rCi_,Fi_,Mi_,Fji_)
% rAi_ is the position vector of Ai
% rAj_ is the position vector of Aj
% rCi_ is the position vector of CM of (i)
% Fi_ total force on (i) at Ci including
%      the inertia force at Ci and
%      the gravitational force at Ci
% Mi_ is the total moment on (i) including
%      the moment of inertia on (i)
% Fji_ is the reaction force of (j) on (i) at Aj

% sum forces of (i)
% FAi_ + Fi_+ Fji_ = 0_ ; =>
% FAi_ rection force on (i) at Ai
FAi_ = - Fi_ - Fji_;

% equilibrium motor
```

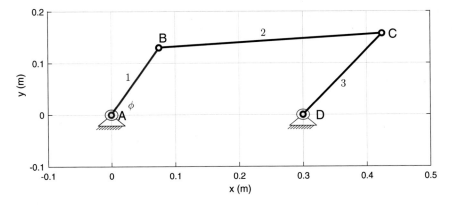

Fig. 6.3 R-RRR mechanism

```
% sum of moments of (i)   about Ai
% AiCi_ x Fi_ + AiAj_ x Fji_ + Mi_ + Mm_ = 0_ =>
% Mm_ is the equilibrium moment or motor moment
Mm_ = -cross(rCi_-rAi_,Fi_)-cross(rAj_-rAi_,Fji_)-Mi_;

out = [FAi_(1), FAi_(2), Mm_(3)];
end
```

6.4.1 R-RRR Mechanism

An R-RRR Mechanism is considered in Fig. 6.3.
   The input data for the four-bar mechanism are:

```
AB=0.15;    % (m) link 1
BC=0.35;    % (m) link 2
CD=0.20;    % (m) link 3
AD=0.30;    % (m) link 0
phi = pi/3; % (rad) angle of link 1 with x-axis
n = 60;     % (rpm) angular speed of link 1 (constant)

omega = pi*n/30; % (rad/s) angular velocity of 1
alpha = 0; % (rad/s^2) angular acceleration of 1

omega1_ = [0 0 omega]; % angular velocity vector of 1
alpha1_ = [0 0 alpha]; % angular acceleration vector of 1
```

The origin of the system is at joint A and:

```
% Joint A (origin)
xA = 0; yA = 0;
rA_ = [xA yA 0]; % position of A
vA_ = [0 0 0];   % velocity of A
aA_ = [0 0 0];   % acceleration of A
```

For the fixed joint D the input data are:

```
% Joint D (fixed)
xD = AD; yD = 0;
rD_ = [xD yD 0]; % position of D
vD_ = [0 0 0];   % velocity of D
aD_ = [0 0 0];   % acceleration of D
```

To calculate the position, velocity, and acceleration of joint B the routine pvaR(rAi_, vAi_, aAi_, li, phi, omega, alpha) is used where:

```
% Joint B
% call pvaR function
% pvaR(rAi_, vAi_, aAi_, li, phi, omega, alpha)
% rAi_ = rA_
% vAi_ = vA_
% aAi_ = aA_
% li = AB
% phi = phi
% omega = omega
% alpha = alpha
```

The results are:

```
pvaB = pvaR(rA_, vA_, aA_, AB, phi, omega, alpha);

xB = pvaB(1); yB = pvaB(2);
rB_ = [xB yB 0]; % position of B

vBx = pvaB(3); vBy = pvaB(4);
vB_ = [vBx vBy 0];   % velocity of B

aBx = pvaB(5); aBy = pvaB(6);
aB_ = [aBx aBy 0]; % acceleration of B

% rB_ = [ 0.075,   0.130, 0] (m)
% vB_ = [-0.816,   0.471, 0] (m/s)
```

```
% aB_ = [-2.961, -5.128, 0] (m/s^2)
```

The position of joint C is calculated with the help of the function for position analysis for RRR dyad:

```
% position of joint C
% call function pRRR
% posRRR(rAi_, rAj_, li, lj)
% rAi_ = rB_
% rAj_ = rD_
% li = BC
% lj = CD
posC = posRRR(rB_, rD_, BC, CD);
```

The outputs are the two possible solutions for the joint C and the correct solution can be selected with the conditions: $y_C > y_B$:

```
xCI = posC(1); yCI = posC(2);
xCII = posC(3); yCII = posC(4);
% select correct C
if yCI > yB xC = xCI; yC = yCI;
else xC = xCII; yC = yCII; end
rC_ = [xC yC 0]; % position vector of C
% rC_ = [ 0.424,  0.157, 0] (m)
```

The angles of the links with the horizontal axis are:

```
% angle of link 2 with x-axis
phi2 = atan((yB-yC)/(xB-xC));
% angle of link 3 with x-axis
phi3 = atan((yC-yD)/(xC-xD));
% phi2 =  4.433 (degrees)
% phi3 = 51.701 (degrees)
```

The velocity and the acceleration of C are calculated with the velaccRRR.m routine:

```
% velaccRRR(rAi_,rAj_,rP_,vAi_,vAj_,aAi_,aAj_)
% rAi_ = rB_
% rAj_ = rD_
% rP_  = rC_
% vAi_ = vB_
% vAj_ = vD_
% aAi_ = aB_
% aAj_ = aD_
```

```
vaC = velaccRRR(rB_,rD_,rC_,vB_,vD_,aB_,aD_);
vCx = vaC(1);
vCy = vaC(2);
vC_ = [vCx vCy 0]; % velocity vector of C
aCx = vaC(3);
aCy = vaC(4);
aC_ = [aCx aCy 0]; % acceleration vector of C
% vC_ = [-0.831,   0.656,  0] (m/s)
% aC_ = [-3.093,  -4.693,  0] (m/s^2)
```

The angular velocity and acceleration of links 2 and 3 are obtained with
angvelacc.m function:

```
% angular velocity acceleration link 2
% call   angvelacc
% angvelacc(rP_, rQ_, vP_, vQ_, aP_, aQ_)
% rP_ = rB_
% rQ_ = rC_
% vP_ = vB_
% vQ_ = vC_
% aP_ = aB_
% aQ_ = aC_
angvelacc2 = angvelacc(rB_,rC_,vB_,vC_,aB_,aC_);
omega2z = angvelacc2(1);
alpha2z = angvelacc2(2);
omega2_ = [0 0 omega2z];
alpha2_ = [0 0 alpha2z];

% angular velocity and acceleration of link 3
% call   angvelacc
% angvelacc(rP_, rQ_, vP_, vQ_, aP_, aQ_)
% rP_ = rD_
% rQ_ = rC_
% vP_ = vD_
% vQ_ = vC_
% aP_ = aD_
% aQ_ = aC_
angvelacc3 = angvelacc(rD_,rC_,vD_,vC_,aD_,aC_);
omega3z = angvelacc3(1);
alpha3z = angvelacc3(2);
omega3_ = [0 0 omega3z];
alpha3_ = [0 0 alpha3z];

% omega2_ = [0,0,  0.529](rad/s)
% alpha2_ = [0,0,  1.270](rad/s^2)
%
% omega3_ = [0,0,  5.291](rad/s)
% alpha3_ = [0,0, -2.406](rad/s^2)
```

The position vectors and the linear accelerations of the centers of mass (CM) of links 1, 2, and 3 are:

```
rC1_ = rB_/2;          % position vector of CM of 1
rC2_ = (rB_+rC_)/2;    % position vector of CM of 2
rC3_ = (rC_+rD_)/2;    % position vector of CM of 3
aC1_ = aB_/2;          % acceleration of C1
aC2_ = (aB_+aC_)/2;    % acceleration of C2
aC3_ = (aC_+aD_)/2;    % acceleration of C3
% rC1_ = [ 0.038, 0.065,0]  (m)
% rC2_ = [ 0.249, 0.143,0]  (m)
% rC3_ = [ 0.362, 0.078,0]  (m)
% aC1_ = [-1.480,-2.564,0]  (m/s^2)
% aC2_ = [-3.027,-4.911,0]  (m/s^2)
% aC3_ = [-1.546,-2.346,0]  (m/s^2)
```

The masses m_i, the forces of gravity $\mathbf{G}_i$, the forces of inertia $\mathbf{F}_{\mathrm{in}i}$, and the moments of inertia $\mathbf{M}_{\mathrm{in}i}$ are calculated for each link $i = 1, 2, 3$:

```
h = 0.01;     % (m) height of the links
d = 0.001;    % (m) depth of the links
rho = 8000;   % (kg/m^3) density of the material
g = 9.807;    % (m/s^2) gravitational acceleration

fprintf('Link 1 \n')
m1 = rho*AB*h*d;         % (kg) mass
G1_ = [0,-m1*g,0];       % (N) force of gravity
Fin1_ = -m1*aC1_;        % (N) force of inertia
IC1 = m1*(AB^2+h^2)/12;  % (kg m^2) mass moment of inertia about C1
IA = IC1 + m1*(AB/2)^2;  % (kg m^2) mass moment of inertia about A
Min1_ = -IC1*alpha1_;    % (N m) moment of inertia

fprintf('Link 2 \n')
m2 = rho*BC*h*d;         % (kg) mass
G2_ = [0,-m2*g,0];       % (N) force of gravity
Fin2_ = -m2*aC2_;        % (N) force of inertia
IC2 = m2*(BC^2+h^2)/12;  % (kg m^2) mass moment of inertia about C2
Min2_ = -IC2*alpha2_;    % (N m) moment of inertia

fprintf('Link 3 \n')
m3 = rho*CD*h*d;         % (kg) mass
G3_ = [0,-m3*g,0];       % (N) force of gravity
Fin3_ = -m3*aC3_;        % (N) force of inertia
IC3 = m3*(CD^2+h^2)/12;  % (kg m^2) mass moment of inertia about C3
IC = IC3+m3*(CD/2)^2;    % (kg m^2) mass moment of inertia about C
Min3_ = -IC3*alpha3_;    % (N m) moment of inertia

% Link 1
% m1 = 0.012 (kg)
% G1_ = - m1 g_ = [0,-0.118,0] (N)
```

```
% Fin1_= - m1 aC1_ = [ 0.018, 0.031,0] (N)
% IC1 = 2.26e-05 (kg m^2)
% IA = 9.01e-05 (kg m^2)
% Min1_=-IC1 alpha1_=[0, 0, 0](N m)
%
% Link 2
% m2 =  0.028 (kg)
% G2_ = - m2 g_ = [0,-0.275,0] (N)
% Fin2_= - m2 aC2 =[ 0.085, 0.137,0] (N)
% IC2 = 2.861e-04 (kg m^2)
% Min2_=-IC2 alpha2=[0,0,-3.633e-04](N m)
%
% Link 3
% m3 =  0.016 (kg)
% G3_ = - m3 g_ = [0, -0.157, 0] (N)
% Fin3_ = - m3 aC3 =[ 0.025, 0.038,0] (N)
% IC3 = 5.347e-05 (kg m^2)
% Min3_=-IC3 alpha3=[0,0,1.286e-04](N m)
```

There is an external moment on the last link 3:

```
% external moment on 3
Me = 100; % (N m)
Me_ = -sign(omega3_(3))*[0,0,Me];
% Me_ = [0, 0, -100.000] (N m)
```

The total external forces on the dyad RRR (links 2 and 3) are:

```
F2_ = Fin2_ + G2_; % total force on 2 at C2
M2_ = Min2_;       % total moment on 2
F3_ = Fin3_ + G3_; % total force on 3 at C3
M3_ = Min3_+ Me_;  % total moment on 3
```

The joint reaction forces on the dyad are determined with the dforceRRR.m routine:

```
% call dforceRRR
% dforceRRR(rAi_,rAj_,rP_,rCi_,rCj_,Fi_,Mi_,Fj_,Mj_)
% out=[FAixs,FAiys,FAjxs,FAjys,Fjix,Fjiy]
% rAi_ = rB_
% rAj_ = rD_
% rP_  = rC_
% rCi_ = rC2_
% rCj_ = rC3_
% Fi_  = F2_
% Mi_  = M2_
% Fj_  = F3_
% Mj_  = M3_
```

and the results are:

```
FD23_ = dforceRRR(rB_,rD_,rC_,rC2_,rC3_,F2_,M2_,F3_,M3_);

F12x = FD23_(1);
F12y = FD23_(2);
F03x = FD23_(3);
F03y = FD23_(4);
F32x = FD23_(5);
F32y = FD23_(6);

F12_ = [F12x F12y 0]; % reaction force of 1 on 2 at B
F03_ = [F03x F03y 0]; % reaction force of 0 on 3 at D
F32_ = [F32x F32y 0]; % reaction force of 3 on 2 at C

% F12_ = [-678.875,-52.561,0] (N)
% F03_ = [678.766,52.818,0] (N)
% F32_ = [678.791,52.699,0] (N)
```

For the driver link 1 the forces are calculated with dforceR.m:

```
% dforceR(rAi_,rAj_,rCi_,Fi_,Mi_,Fji_)
% rAi_ = rA_
% rAj_ = rB_
% rCi_ = rC1_
% Fi_  = F1
% Mi_  = M1
% Fji_ = F21_=-F12
F1_ = Fin1_ + G1_; % total force on 1
M1_ = Min1_;       % total moment on 1
FD1_ = dforceR(rA_,rB_,rC1_,F1_,M1_,-F12_);
F01x = FD1_(1);
F01y = FD1_(2);
Mm = FD1_(3);
F01_ = [F01x F01y 0]; % reaction force of 0 on 1 at A
Mm_ = [0 0 Mm];  % equilibrium or motor moment on 1

% F01_ = [-678.893,-52.474,0] (N)
% Mm_  = [0,0,84.251] (N m)
```

Fig. 6.4 R-RTR mechanism

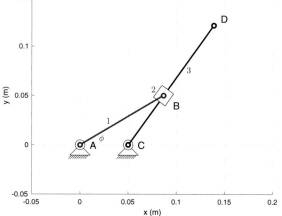

6.4.2   R-RTR Mechanism

Figure 6.4 shows an R-RTR mechanism.
   The input data for the mechanism are:

```
AB = 0.10 ;   % (m) link 1
% link 2 is the slider at B
AC = 0.05 ;   % (m) link 0
CD = 0.15 ;   % (m) link 3

phi = pi/6;   % (rad) angle of link 1 with x-axis
n = -50;      % (rpm) angular speed of link 1 (constant)

omega = pi*n/30; % (rad/s) angular velocity of 1
alpha = 0; % (rad/s^2) angular acceleration of 1

omega1_ = [0 0 omega]; % angular velocity vector of 1
alpha1_ = [0 0 alpha]; % angular acceleration vector of 1

% Joint A (origin)
xA = 0; yA = 0;
rA_ = [xA yA 0]; % position of A
vA_ = [0 0 0];   % velocity of A
aA_ = [0 0 0];   % acceleration of A

% Joint C (fixed)
xC = AC ; yC = 0;
rC_ = [xC yC 0]; % position of C
```

```
vC_ = [0 0 0];   % velocity of C
aC_ = [0 0 0];   % acceleration of C
```

For the point B on links 1 and 2 ($B = B_1 = B_2$) the position, velocity, and acceleration are determined with pvaR.m:

```
% point B=B1=B2
% pvaR(rAi_, vAi_, aAi_, li, phi, omega, alpha)
% rAi_ = rA_
% vAi_ = vA_
% aAi_ = aA_
% li = AB
% phi = phi
% omega = omega
% alpha = alpha
pvaB = pvaR(rA_, vA_, aA_, AB, phi, omega, alpha);

xB = pvaB(1); yB = pvaB(2);
rB_ = [xB yB 0]; % position of B

vBx = pvaB(3); vBy = pvaB(4);
vB_ = [vBx vBy 0];   % velocity of B

aBx = pvaB(5); aBy = pvaB(6);
aB_ = [aBx aBy 0]; % acceleration of B

% rB_ = [ 0.087, 0.050,0]  (m)
% vB_ = [ 0.262, -0.453, 0]  (m/s)
% aB_ = [-2.374, -1.371, 0]  (m/s^2)
```

The angles of the links 2 and 3 with the horizontal are:

```
% angle of link 2 with x-axis
phi2 = atan((yB-yC)/(xB-xC));
% angle of link 3 with x-axis
phi3 = phi2;
% phi2 = phi3 = 53.794 (degrees)
```

The position of point D is:

```
% call pos3P function
% C, B, and D same direction
% pos3P(rP_, rQ_ , PR)
% rP_ = rC_
% rQ_ = rB_
```

```
% PR = CD: C->B->D
posD = pos3P(rC_, rB_, CD);
xD = posD(1);
yD = posD(2);
rD_ = [xD yD 0];
% rD_ = [ 0.139, 0.121,0] (m)
```

The angular velocity and acceleration of links 2 and 3 are determinated with:

```
% angular velocity and acceleration of link 2
% call   angvelacc
% angvelacc(rP_,rQ_, vP_,vQ_, aP_,aQ_)
% rP_ = rB_
% rQ_ = rC_
% vP_ = vB_
% vQ_ = vC_
% aP_ = aB_
% aQ_ = aC_
angvelacc2 = angvelacc(rB_,rC_,vB_,vC_,aB_,aC_);
omega2z = angvelacc2(1);
alpha2z = angvelacc2(2);
omega2_ = [0 0 omega2z];   % angular velocity of link 2
alpha2_ = [0 0 alpha2z];   % angular acceleration of link 2
omega3_ = omega2_; % angular velocity of link 3
alpha3_ = alpha2_; % angular acceleration of link 3
% omega2_ = omega3_ = [0,0,-7.732](rad/s)
% alpha2_ = alpha3_ = [0,0,-34.865](rad/s^2)
```

The routine linvelacc.m is applied to calculate the velocity and acceleration of point D:

```
% linvelacc(rP_,rQ_,vP_,aP_,omega,alpha)
% rP_ = rC_
% rQ_ = rD_
% vP_ = vC_
% aP_ = aC_
% omega = omega2z
% alpha = alpha2z
vaD = linvelacc(rC_,rD_,vC_,aC_,omega2z,alpha2z);
vDx = vaD(1);
vDy = vaD(2);
aDx = vaD(3);
aDy = vaD(4);
vD_ = [vDx vDy 0]; % velocity of D
aD_ = [aDx aDy 0]; % acceleration of D
```

```
% vD_  = [ 0.936,-0.685,0]  (m/s)
% aD_  = [-1.077,-10.324,0]  (m/s^2)
```

The positions and accelerations of the centers of mass of the links are:

```
rC1_  = rB_/2;        % position vector of CM of 1
rC2_  = rB_;          % position vector of CM of 2
rC3_  = (rD_+rC_)/2;  % position vector of CM of 3
aC1_  = aB_/2;        % acceleration of C1
aC2_  = aB_;          % acceleration of C2
aC3_  = (aD_+aC_)/2;  % acceleration of C3
% rC1_  = [ 0.043, 0.025,0]  (m)
% rC2_  = [ 0.087, 0.050,0]  (m)
% rC3_  = [ 0.094, 0.061,0]  (m)
% aC1_  = [-1.187,-0.685,0]  (m/s^2)
% aC2_  = [-2.374,-1.371,0]  (m/s^2)
% aC3_  = [-0.538,-5.162,0]  (m/s^2)
```

The inertia forces and moments for each link are:

```
h = 0.01;         % (m) height of the links
d = 0.001;        % (m) depth of the links
hSlider = 0.02;   % (m) height of the slider
wSlider = 0.04;   % (m) depth of the slider
rho = 8000;       % (kg/m^3) density of the material
g = 9.807;        % (m/s^2) gravitational acceleration

fprintf('Link 1 \n')
m1 = rho*AB*h*d;        % (kg) mass
G1_ = [0,-m1*g,0];      % (N) force of gravity
Fin1_ = -m1*aC1_;       % (N) force of inertia
IC1 = m1*(AB^2+h^2)/12; % (kg m^2) mass moment of inertia about C1
IA = IC1 + m1*(AB/2)^2; % (kg m^2) mass moment of inertia about A
Min1_ = -IC1*alpha1_;   % (N m) moment of inertia

fprintf('Link 2 \n')
m2 = rho*hSlider*wSlider*d; % (kg) mass
G2_ = [0,-m2*g,0];          % (N) force of gravity
Fin2_ = -m2*aC2_;           % (N) force of inertia
IC2 = m2*(hSlider^2+wSlider^2)/12; % (kg m^2) mass moment of inertia about C2
Min2_ = -IC2*alpha2_;       % (N m) moment of inertia

fprintf('Link 3 \n')
m3 = rho*CD*h*d;        % (kg) mass
G3_ = [0,-m3*g,0];      % (N) force of gravity
Fin3_ = -m3*aC3_;       % (N) force of inertia
IC3 = m3*(CD^2+h^2)/12; % (kg m^2) mass moment of inertia about C3
ID = IC3+m3*(CD/2)^2;   % (kg m^2) mass moment of inertia about D
Min3_ = -IC3*alpha3_;   % (N m) moment of inertia

% Link 1
% m1 =  0.008 (kg)
% G1_ = - m1 g_ = [0,-0.078,0] (N)
% Fin1_= - m1 aC1_ = [ 0.009, 0.005,0] (N)
```

```
% IC1 = 6.73333e-06 (kg m^2)
% IA = 2.67333e-05 (kg m^2)
% Min1_=-IC1 alpha1_=[0, 0, 0](N m)
%
% Link 2
% m2 =  0.006 (kg)
% G2_ = - m2 g_ = [0,-0.063,0] (N)
% Fin2_= - m2 aC2 =[ 0.015, 0.009,0] (N)
% IC2 = 1.067e-06 (kg m^2)
% Min2_=-IC2 alpha2=[0,0,3.719e-05](N m)
%
% Link 3
% m3 =  0.012 (kg)
% G3_ = - m3 g_ = [0, -0.118, 0] (N)
% Fin3_ = - m3 aC3 =[ 0.006, 0.062,0] (N)
% IC3 = 2.260e-05 (kg m^2)
% Min3_=-IC3 alpha3=[0,0,7.880e-04](N m)
```

The external moment on last link 3 is:

```
Me = 100; % (N m)
Me_ = -sign(omega3_(3))*[0,0,Me];
% Me_ = [0, 0, 100.000] (N m)
```

For the dyad RTR (links 2 and 3) the total forces and moments are:

```
F2_ = Fin2_ + G2_; % total force on 2
M2_ = Min2_;       % total moment on 2
F3_ = Fin3_ + G3_; % total force on 3
M3_ = Min3_+ Me_;  % total moment on 3
```

The joint forces for this dyad are obtained with the routine dforceRTR.m:

```
% dforceRTR(rAi_,rAj_,rCi_,Fi_,Mi_,Fj_,Mj_)
% i=(3)  j=(2)
% rAi_ = rC_
% rAj_ = rB_
% rCi_ = rC3_
% Fi_  = F3_
% Mi_  = M3_
% Fj_  = F2_
% Mj_  = M2_

FD23=dforceRTR(rC_,rB_,rC3_,F3_,M3_,F2_,M2_);
F03x = FD23(1);
F03y = FD23(2);
F12x = FD23(3);
F12y = FD23(4);
F32x = FD23(5);
```

```
F32y = FD23(6);
M32z = FD23(7);

F12_ = [F12x F12y 0]; % reaction force of 1 on 2 at B
F03_ = [F03x F03y 0]; % reaction force of 0 on 3 at C
F32_ = [F32x F32y 0]; % reaction force of 3 on 2 at B
M32_ = [0 0 M32z];    % reaction moment of 3 on 2

% F03_ = [-1302.149,953.291,0] (N)
% F12_ = [1302.128,-953.181,0] (N)
% F32_ = [-1302.143,953.235,0] (N)
% M32_ = [0,0,-0.000037] (N m)
```

For the driver link, the function dforceR.m is exercised to find the reaction of the
ground on link 1 and the equilibrium (motor) moment:

```
F1_ = Fin1_ + G1_; % total force on 1
M1_ = Min1_;       % total moment on 1
FD1_ = dforceR(rA_,rB_,rC1_,F1_,M1_,-F12_);
F01x = FD1_(1);
F01y = FD1_(2);
Mm = FD1_(3);
F01_ = [F01x F01y 0];  % reaction force of 0 on 1 at A
Mm_ = [0 0 Mm]; % equilibrium (motor) moment
% F01_ = [1302.118,-953.108,0] (N)
% Mm_ = [0,0,-147.651] (N m)
```

6.4.3 R-RRT-RTR Mechanism

An R-RRT-RTR mechanism is considered in Fig. 6.5.
The input data are:

```
AB = 0.20 ; % (m) link 1
BC = 0.21 ; % (m) link 2
CD = 0.39 ; % (m) link 2
% link 3 is the slider at D
CF = 0.45 ; % (m) link 4
% link 5 is the slider at E

a = 0.30 ; % (m) distance from E to y-axis
b = 0.25 ; % (m) distance from E to x-axis
c = 0.05 ; % (m) distance from D to x-axis
```

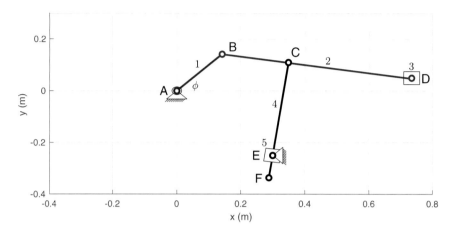

Fig. 6.5 R-RRT-RTR mechanism

```
% Joint A (origin)
xA = 0; yA = 0;
rA_ = [xA yA 0]; % position vector of A
vA_ = [0 0 0];   % velocity vector of A
aA_ = [0 0 0];   % acceleration vector of A

% Joint E (fixed point)
xE =  a;
yE = -b;
rE_ = [xE yE 0]; % position vector of E
vE_ = [0 0 0];   % velocity vector of E
aE_ = [0 0 0];   % acceleration vector of E

% y-coordinate of D
yD = c;

phi = 45*pi/180; % (rad) angle of link 1 with x-axis
n = -500; % (rpm) angular speed of link 1 (constant)
omega = pi*n/30; % (rad/s) angular velocity of 1
alpha = 0; % (rad/s^2) angular acceleration of 1
omega1_ = [0 0 omega]; % angular velocity vector of 1
alpha1_ = [0 0 alpha]; % angular acceleration vector of 1
```

To calculate the position, velocity, and acceleration of B, pvaR.m is used:

```
% call pvaR function
% pvaR(rAi_, vAi_, aAi_, li, phi, omega, alpha)
% rAi_ = rA_
```

```
% vAi_ = vA_
% aAi_ = aA_
% li = AB
% phi = phi
% omega = omega
% alpha = alpha
pvaB = pvaR(rA_, vA_, aA_, AB, phi, omega, alpha);

xB = pvaB(1); yB = pvaB(2);
rB_ = [xB yB 0]; % position of B

vBx = pvaB(3); vBy = pvaB(4);
vB_ = [vBx vBy 0];   % velocity of B

aBx = pvaB(5); aBy = pvaB(6);
aB_ = [aBx aBy 0]; % acceleration of B

% rB_ = [ 0.141, 0.141,0] (m)
% vB_ = [ 7.405,-7.405,0] (m/s)
% aB_ = [-387.715,-387.715,0] (m/s^2)
```

The position of *D* is found with posRRT.m function:

```
% angle of the sliding direction
phi3 = pi;
% As point on the sliding direction
xAs = 0; yAs = c;
rAs_ = [xAs yAs 0];
% call function posRRT
% posRRT(rAi_, rAs_, li, Theta)
% rAi_ = rB_
% rAs_ = rAs_ = [0 c 0]  As on the x-sliding direction
% li = BC+CD
% Theta = phi3 angle of the sliding direction with x-axis
posD = posRRT(rB_, rAs_, BC+CD, phi3);

xD1 = posD(1);
xD2 = posD(3);

% select the correct position of D
% 0 < phi < pi/2
if xD1 > xB xD = xD1; else xD = xD2; end
rD_ = [xD yD 0]; % position vector of D
% rD_ = [ 0.734, 0.05, 0] (m)
```

The angle of link 2 is:

```
% angle of link 2 with x-axis
phi2 = atan2((yB-yD),(xB-xD));
% phi2 = 171.236 (degrees)
```

The routine pos3P.m calculates the position of C:

```
% call function pos3P (3 points on a line)
% pos3P(rP_, rQ_ , PR)
% rP_ = rB_
% rQ_ = rD_
% PR = BC: B->C->D
posC = pos3P(rB_, rD_, BC);

xC = posC(1);
yC = posC(2);
rC_ = [xC yC 0]; % position vector of C
% rC_ = [ 0.349,  0.109, 0] (m)
```

The angle of links 4 and 5 is:

```
% angle of link 4 (link 5) with x-axis
phi4 = atan2((yC-yE),(xC-xE));
phi5 = phi4;
% phi4 = phi5 = 82.242 (degrees)
```

The position of F is:

```
xF = xC-CF*cos(phi4);
yF = yC-CF*sin(phi4);
rF_ = [xF yF 0];
% rF_ = [ 0.288, -0.336, 0] (m)
```

The velocity and acceleration of D are:

```
% As a point on the sliding direction
% xAs = 0; yAs = c; => rAs_ = [xAs yAs 0];
vAs_ = [0 0 0]; % velocity of As
aAs_ = [0 0 0]; % acceleration of As

% velaccRRT(rAi_,rAs_,rAj_,vAi_,vAs_,aAi_,aAs_,dtheta,ddtheta)
% rAi_ = rB_
% rAs_ = rAs_
% rAj_ = rD_
```

```
% vAi_  = vB_
% vAs_  = vAs_
% aAi_  = aB_
% aAs_  = aAs_
% dtheta = 0 slider direction is fixed
% ddtheta = 0
vaD=velaccRRT(rB_,rAs_,rD_,vB_,vAs_,aB_,aAs_,0,0);
vDx = vaD(1);
vDy = vaD(2);
vD_  = [vDx vDy 0]; % velocity vector of D
aDx = vaD(3);
aDy = vaD(4);
aD_  = [aDx aDy 0]; % acceleration vector of D

% vD_ = [ 8.546, 0, 0] (m/s)
% aD_ = [-422.604, 0, 0] (m/s^2)
```

The angular velocity and acceleration of link 2 are:

```
% call  angvelacc
% angvelacc(rP_,rQ_, vP_,vQ_, aP_,aQ_)
% rP_  = rB_
% rQ_  = rD_
% vP_  = vB_
% vQ_  = vD_
% aP_  = aB_
% aQ_  = aD_
ang2=angvelacc(rB_,rD_,vB_,vD_,aB_,aD_);
omega2z = ang2(1);
alpha2z = ang2(2);

% omega2_ = [0,0,12.487](rad/s)
% alpha2_ = [0,0,629.786](rad/s^2)
```

The angular velocity and acceleration of slider 3 are:

```
omega3_ = [0 0 0];
alpha3_ = [0 0 0];
```

The velocity and acceleration of C are:

```
% linvelacc(rP_,rQ_,vP_,aP_,omega,alpha)
% rP_  = rB_
% rQ_  = rC_
% vP_  = vB_
```

```
% aP_ = aB_
% omega = omega2z
% alpha = alpha2z
vaC = linvelacc(rB_,rC_,vB_,aB_,omega2z,alpha2z);
vCx = vaC(1);
vCy = vaC(2);
vC_ = [vCx vCy 0]; % velocity vector of C
aCx = vaC(3);
aCy = vaC(4);
aC_ = [aCx aCy 0]; % acceleration vector of C

% vC_ = [ 7.804, -4.813, 0] (m/s)
% aC_ = [-399.926, -252.015, 0] (m/s^2)
```

The angular velocity and acceleration of link 4 and slider 5 are:

```
% call   angvelacc
% angvelacc(rP_,rQ_, vP_,vQ_, aP_,aQ_)
% rP_ = rC_
% rQ_ = rE_
% vP_ = vC_
% vQ_ = vE_
% aP_ = aC_
% aQ_ = aE_
ang4=angvelacc(rC_,rE_,vC_,vE_,aC_,aE_);
omega4z = ang4(1);
alpha4z = ang4(2);

omega4_ = [0 0 omega4z];  % angular velocity of 4
alpha4_ = [0 0 alpha4z];  % angular acceleration of 4

omega5_ = omega4_;  % angular velocity of 5
alpha5_ = alpha4_;  % angular acceleration of 5

% omega4_ = omega5_ = [0,0,-23.109](rad/s)
% alpha4_ = alpha5_ = [0,0,525.219](rad/s^2)
```

The velocity and acceleration of F are:

```
% linvelacc(rP_,rQ_,vP_,aP_,omega,alpha)
% rP_ = rC_
% rQ_ = rF_
% vP_ = vC_
% aP_ = aC_
% omega = omega4z
```

```
% alpha = alpha4z
vaF=linvelacc(rC_,rF_,vC_,aC_,omega4z,alpha4z);
vFx = vaF(1);
vFy = vaF(2);
vF_ = [vFx vFy 0]; % velocity vector of F
aFx = vaF(3);
aFy = vaF(4);
aF_ = [aFx aFy 0]; % acceleration vector of F

% vF_ = [-2.500, -3.409, 0] (m/s)
% aF_ = [-133.299, -45.808, 0] (m/s^2)
```

The position vectors and the acceleration vectors of the centers of mass of the links are:

```
rC1_ = rB_/2;         % position vector of CM of 1
rC2_ = (rB_+rD_)/2;   % position vector of CM of 2
rC3_ = rD_;           % position vector of CM of 3
rC4_ = (rC_+rF_)/2;   % position vector of CM of 4
rC5_ = rE_;           % position vector of CM of 5

aC1_ = aB_/2;         % acceleration of C1
aC2_ = (aB_+aD_)/2;   % acceleration of C2
aC3_ = aD_;           % acceleration of C3
aC4_ = (aC_+aF_)/2;   % acceleration of C4
aC5_ = [0 0 0];       % acceleration of C5
```

The external moment on link 5 is:

```
Me_ = -sign(omega5_(3))*[0 0 1000];
% Me_ = [0, 0, 1000.000] (N m)
```

The inertia forces and inertia moments on the links are:

```
h = 0.02;          % (m) height of the links
d = 0.01;          % (m) depth of the links
hSlider = 0.04;    % (m) height of the sliders
wSlider = 0.08;    % (m) width of the sliders
dSlider = 0.02;    % (m) depth of the sliders
rho = 8000;        % (kg/m^3) density of the material
g = 9.807;         % (m/s^2) gravitational acceleration

fprintf('Link 1 \n')
m1 = rho*AB*h*d;
Fin1_ = -m1*aC1_;
```

```
G1_ = [0,-m1*g,0];
IC1 = m1*AB^2/12;
Min1_ = -IC1*alpha1_;

fprintf('Link 2 \n')
m2 = rho*(BC+CD)*h*d;
Fin2_ = -m2*aC2_;
G2_ = [0,-m2*g,0];
IC2 = m2*(BC+CD)^2/12;
Min2_ = -IC2*alpha2_;

fprintf('Link 3 \n')
m3 = rho*hSlider*wSlider*dSlider;
Fin3_ = -m3*aC3_;
G3_ = [0,-m3*g,0];
IC3 = m3*(hSlider^2+wSlider^2)/12;
Min3_ = -IC3*alpha3_;

fprintf('Link 4 \n')
m4 = rho*CF*h*d;
Fin4_ = -m4*aC4_;
G4_ = [0,-m4*g,0];
IC4 = m4*CF^2/12;
Min4_ = -IC4*alpha4_;

fprintf('Link 5 \n')
m5 = rho*hSlider*wSlider*dSlider;
Fin5_ = -m5*aC5_;
G5_ = [0,-m5*g,0];
IC5 = m5*(hSlider^2+wSlider^2)/12;
Min5_ = -IC5*alpha5_;

% Link 1
% m1 =  0.320 (kg)
% Fin1_= - m1 aC1_ = [62.034,62.034,0] (N)
% G1_ = - m1 g = [0,-3.138,0] (N)
% IC1 = 0.00106667 (kg m^2)
% Min1_=-IC1 alpha1_=[0, 0, 0](N m)
%
% Link 2
% m2 =  0.960 (kg)
% Fin2_= - m2 aC2_ = [388.953,186.103,0] (N)
% G2_ = - m2 g = [0,-9.415,0] (N)
% IC2 = 0.0288 (kg m^2)
% Min2_=-IC2 alpha2_=[0, 0, -1.813784e+01](N m)
```

```
%
% Link 3
% m3 =  0.512 (kg)
% Fin3_= - m3 aC3_ =[216.373,-0.000,0] (N)
% G3_ = - m3 g_ = [0,-5.021,0] (N)
% IC3 = 3.413e-04 (kg m^2)
% Min3_=-IC3 alpha3_=[0,0,    000](N m)
%
% Link 4
% m4 =  0.720 (kg)
% Fin4_= - m4 aC4_ =[191.961,107.216,0] (N)
% G4_ = - m4 g_ = [0,-7.061,0] (N)
% IC4 = 1.215e-02 (kg m^2)
% Min4_=-IC4 alpha4_=[0,0,-6.381e+00](N m)
%
% Link 5
% m5 =  0.512 (kg)
% Fin5_ = - m5 aC5 =[-0.000,-0.000,0] (N)
% G5_ = - m5 g = [0, -5.021, 0] (N)
% IC5 = 3.413e-04 (kg m^2)
% Min5_=-IC5 alpha5_=[0, 0, -1.793e-01](N m)
```

For force analysis the analysis starts with the last dyad RTR (links 4 and 5). The total forces on the last dyad are:

```
F5_ = Fin5_ + G5_; % total force on 5
M5_ = Min5_ + Me_; % total moment on 5
F4_ = Fin4_ + G4_; % total force on 4
M4_ = Min4_;       % total moment on 4
```

The joint reaction forces for the dyad are computed with:

```
% RTR -> CEE (4 & 5)
% dforceRTR(rAi_,rAj_,rCi_,Fi_,Mi_,Fj_,Mj_)
% out = [FAixs,FAiys,FAjxs,FAjys,Fijx,Fijy,Mijz];
% rAi_ = rC_
% rAj_ = rE_
% rCi_ = rC4_
% Fi_ = F4_
% Mi_ = M4_
% Fj_ = F5_
% Mj_ = M5_

FD45=dforceRTR(rC_,rE_,rC4_,F4_,M4_,F5_,M5_);
F24x = FD45(1);
```

```
F24y = FD45(2);
F05x = FD45(3);
F05y = FD45(4);
F45x = FD45(5);
F45y = FD45(6);
M45z = FD45(7);


F24_ = [F24x F24y 0]; % reaction force of 2 on 4 at C
F05_ = [F05x F05y 0]; % reaction force of 0 on 5 at E
F45_ = [F45x F45y 0]; % reaction force of 4 on 5 at E
M45_ = [0 0 M45z];     % reaction moment of 4 on 5 at E

% F05_ = [-2822.194,389.528,0] (N)
% F24_ = [2630.233,-484.662,0] (N)
% F45_ = [2822.194,-384.507,0] (N)
% M45_ = [0,0,-999.821] (N m)
```

For the dyad RRT (links 2 and 3) the total forces:

```
F2_ = Fin2_ + G2_ - F24_;  % total force on 2
M2_ = Min2_+cross(rC_-rC2_,-F24_); % total moment on 2
F3_ = Fin3_ + G3_; % total force on 3
M3_ = Min3_; % total moment on 3
```

The joint reaction forces for the RRT dyad are evaluated with:

```
% dforceRRT(rAi_,rAs_,rAj_,rCi_,Fi_,Mi_,Fj_,Mj_)
% out = [FAixs,FAiys,FAjxs,FAjys,Fijx,Fijy,MAjz];
% rAi_ = rB_
% rAs_ = rAs_
% rAj_ = rD_
% rCi_ = rC2_
% Fi_  = F2_
% Mi_  = M2_
% Fj_  = F3_
% Mj_  = M3_
FD23=dforceRRT(rB_,rAs_,rD_,rC2_,F2_,M2_,F3_,M3_);
F12xs = FD23(1);
F12ys = FD23(2);
F03xs = FD23(3);
F03ys = FD23(4);
F23xs = FD23(5);
F23ys = FD23(6);
M03zs = FD23(7);
```

```
F12_ = [F12xs F12ys 0];  % reaction force of 1 on 2 at B
F03_ = [F03xs F03ys 0];  % reaction force of 0 on 3 at D
F23_ = [F23xs F23ys 0];  % reaction force of 2 on 3 at D
M03_ = [0 0 M03zs];      % reaction moment of 0 on 3

% F03_ = [ 0.000,-143.783,0]  (N)
% F12_ = [2024.907,-512.546,0]  (N)
% F23_ = [-216.373,148.804,0]  (N)
% M03_ = [0,0,0]  (N m)
```

The joint reaction force of the ground on link 1 and the motor moment on link 1 are calculated with:

```
F21_ = -F12_;      % joint force of 2 on 1 at B
F1_  = Fin1_+G1_;  % total force on 1
M1_  = Min1_;      % total moment on 1
FDL = dforceR(rA_,rB_,rC1_,F1_,M1_,F21_);

F01xs = FDL(1);
F01ys = FDL(2);
Mms = FDL(3);

F01_ = [F01xs F01ys 0];  % reaction force of 0 on 1
Mm_  = [0 0 Mms];  % equilibrium (motor) moment

% F01_ = [1962.873,-571.442,0]  (N)
% Mm_  = [0,0,-358.628]  (N m)
```

6.4.4 Problems

6.1 Restate Problem 4.1 using the dyad routines. Compare the results.

6.2 Repeat Problem 4.2 using the dyad routines. Compare with the results obtained with the classic method.

6.3 Find the kinematics and the dynamics of the mechanism in Problem 4.3 using the dyad routines. Compare with the results obtained with the classic method.

6.4 Repeat Problem 4.4 using the dyad routines. Compare with the results obtained with the classic method.

6.5 Redo Problem 4.5 using the dyad routines. Compare with the results obtained with the classic method.

References

1. E.A. Avallone, T. Baumeister, A. Sadegh, *Marks' Standard Handbook for Mechanical Engineers*, 11th edn. (McGraw-Hill Education, New York, 2007)
2. I.I. Artobolevski, *Mechanisms in Modern Engineering Design* (MIR, Moscow, 1977)
3. M. Atanasiu, *Mechanics* (Mecanica), EDP, Bucharest, 1973)
4. H. Baruh, *Analytical Dynamics* (WCB/McGraw-Hill, Boston, 1999)
5. A. Bedford, W. Fowler, *Dynamics* (Addison Wesley, Menlo Park, CA, 1999)
6. A. Bedford, W. Fowler, *Statics* (Addison Wesley, Menlo Park, CA, 1999)
7. F.P. Beer et al., *Vector Mechanics for Engineers: Statics and Dynamics* (McGraw-Hill, New York, NY, 2016)
8. M. Buculei, D. Bagnaru, G. Nanu, D.B. Marghitu, *Computing Methods in the Analysis of the Mechanisms with Bars* (Scrisul Romanesc, Craiova, 1986)
9. M.I. Buculei, *Mechanisms* (University of Craiova Press, Craiova, Romania, 1976)
10. D. Bolcu, S. Rizescu, *Mecanica* (EDP, Bucharest, 2001)
11. J. Billingsley, *Essential of Dynamics and Vibration* (Springer, 2018)
12. R. Budynas, K.J. Nisbett, *Shigley's Mechanical Engineering Design*, 9th edn. (McGraw-Hill, New York, 2013)
13. J.A. Collins, H.R. Busby, and G.H. Staab, *Mechanical Design of Machine Elements and Machines*, 2nd edn (Wiley, 2009)
14. M. Crespo da Silva, *Intermediate Dynamics for Engineers* (McGraw-Hill, New York, 2004)
15. A.G. Erdman, G.N. Sandor, *Mechanisms Design* (Prentice-Hall, Upper Saddle River, NJ, 1984)
16. M. Dupac, D.B. Marghitu, *Engineering Applications: Analytical and Numerical Calculation with MATLAB* (Wiley, Hoboken, NJ, 2021)
17. A. Ertas, J.C. Jones, *The Engineering Design Process* (Wiley, New York, 1996)
18. F. Freudenstein, An application of boolean algebra to the motion of epicyclic drivers. J. Eng. Ind. 176–182 (1971)
19. J.H. Ginsberg, *Advanced Engineering Dynamics* (Cambridge University Press, Cambridge, 1995)
20. H. Goldstein, *Classical Mechanics* (Addison-Wesley, Redwood City, CA, 1989)
21. D.T. Greenwood, *Principles of Dynamics* (Prentice-Hall, Englewood Cliffs, NJ, 1998)
22. A.S. Hall, A.R. Holowenko, H.G. Laughlin, *Schaum's Outline of Machine Design* (McGraw-Hill, New York, 2013)
23. B.G. Hamrock, B. Jacobson, S.R. Schmid, *Fundamentals of Machine Elements* (McGraw-Hill, New York, 1999)
24. R.C. Hibbeler, *Engineering Mechanics: Dynamics* (Prentice Hall, 2010)
25. T.E. Honein, O.M. O'Reilly, On the Gibbs–Appell equations for the dynamics of rigid bodies. J. Appl. Mech. **88**/074501-1 (2021)
26. R.C. Juvinall, K.M. Marshek, *Fundamentals of Machine Component Design*, 5th edn. (Wiley, New York, 2010)
27. T.R. Kane, *Analytical Elements of Mechanics*, vol. 1 (Academic, New York, 1959)
28. T.R. Kane, *Analytical Elements of Mechanics*, vol. 2 (Academic, New York, 1961)
29. T.R. Kane, D.A. Levinson, "The Use of Kane's Dynamical Equations in Robotics". MIT International Journal of Robotics Research, No. **3**, 3–21 (1983)
30. T.R. Kane, P.W. Likins, D.A. Levinson, *Spacecraft Dynamics* (McGraw-Hill, New York, 1983)
31. T.R. Kane, D.A. Levinson, *Dynamics* (McGraw-Hill, New York, 1985)
32. K. Lingaiah, *Machine Design Databook*, 2nd edn. (McGraw-Hill Education, New York, 2003)
33. N.I. Manolescu, F. Kovacs, A. Oranescu, *The Theory of Mechanisms and Machines* (EDP, Bucharest, 1972)
34. D.B. Marghitu, *Mechanical Engineer's Handbook* (Academic, San Diego, CA, 2001)
35. D.B. Marghitu, M.J. Crocker, *Analytical Elements of Mechanisms* (Cambridge University Press, Cambridge, 2001)
36. D.B. Marghitu, *Kinematic Chains and Machine Component Design* (Elsevier, Amsterdam, 2005)

37. D.B. Marghitu, *Mechanisms and Robots Analysis with MATLAB* (Springer, New York, N.Y., 2009)
38. D.B. Marghitu, M. Dupac, *Advanced Dynamics: Analytical and Numerical Calculations with MATLAB* (Springer, New York, N.Y., 2012)
39. D.B. Marghitu, M. Dupac, H.M. Nels, *Statics with MATLAB* (Springer, New York, N.Y., 2013)
40. D.B. Marghitu, D. Cojocaru, *Advances in Robot Design and Intelligent Control* (Springer International Publishing, Cham, Switzerland, 2016), pp. 317–325
41. D.J. McGill, W.W. King, *Engineering Mechanics: Statics and an Introduction to Dynamics* (PWS Publishing Company, Boston, 1995)
42. J.L. Meriam, L.G. Kraige, *Engineering Mechanics: Dynamics* (Wiley, New York, 2007)
43. C.R. Mischke, Prediction of Stochastic Endurance Strength. Transaction of ASME, Journal Vibration, Acoustics, Stress, and Reliability in Design **109**(1), 113–122 (1987)
44. L. Meirovitch, *Methods of Analytical Dynamics* (Dover, 2003)
45. R.L. Mott, *Machine Elements in Mechanical Design* (Prentice Hall, Upper Saddle River, NJ, 1999)
46. W.A. Nash, *Strength of Materials*. Schaum's Outline Series (McGraw-Hill, New York, 1972)
47. R.L. Norton, *Machine Design* (Prentice-Hall, Upper Saddle River, NJ, 1996)
48. R.L. Norton, *Design of Machinery* (McGraw-Hill, New York, 1999)
49. O.M. O'Reilly, *Intermediate Dynamics for Engineers Newton-Euler and Lagrangian Mechanics* (Cambridge University Press, UK, 2020)
50. O.M. O'Reilly, *Engineering Dynamics: A Primer* (Springer, NY, 2010)
51. W.C. Orthwein, *Machine Component Design* (West Publishing Company, St. Paul, 1990)
52. L.A. Pars, *A Treatise on Analytical Dynamics* Wiley, New York, 1965)
53. F. Reuleaux, *The Kinematics of Machinery* (Dover, New York, 1963)
54. D. Planchard, M. Planchard, *SolidWorks 2013 Tutorial with Video Instruction* (SDC Publications, 2013)
55. I. Popescu, *Mechanisms* (University of Craiova Press, Craiova, Romania, 1990)
56. I. Popescu, C. Ungureanu, *Structural Synthesis and Kinematics of Mechanisms with Bars* (Universitaria Press, Craiova, Romania, 2000)
57. I. Popescu, L. Luca, M. Cherciu, D.B. Marghitu, *Mechanisms for Generating Mathematical Curves* (Springer Nature, Switzerland, 2020)
58. C.A. Rubin, *The Student Edition of Working Model* (Addison-Wesley Publishing Company, Reading, MA, 1995)
59. J. Ragan, D.B. Marghitu, Impact of a kinematic link with MATLAB and solidWorks. Appl. Mech. Mater. **430**, 170–177 (2013)
60. J. Ragan, D.B. Marghitu, MATLAB dynamics of a free link with elastic impact, in*International Conference on Mechanical Engineering, ICOME 2013*, May 16–17, 2013, Craiova, Romania (2013)
61. J.C. Samin, P. Fisette, *Symbolic Modeling of Multibody Systems* (Kluwer, 2003)
62. A.A. Shabana, *Computational Dynamics* (Wiley, New York, 2010)
63. I.H. Shames, *Engineering Mechanics - Statics and Dynamics* (Prentice-Hall, Upper Saddle River, NJ, 1997)
64. J.E. Shigley, C.R. Mischke, *Mechanical Engineering Design* (McGraw-Hill, New York, 1989)
65. J.E. Shigley, C.R. Mischke, R.G. Budynas, *Mechanical Engineering Design*, 7th edn. (McGraw-Hill, New York, 2004)
66. J.E. Shigley, J.J. Uicker, *Theory of Machines and Mechanisms* (McGraw-Hill, New York, 1995)
67. D. Smith, *Engineering Computation with MATLAB* (Pearson Education, Upper Saddle River, NJ, 2008)
68. R.W. Soutas-Little, D.J. Inman, *Engineering Mechanics: Statics and Dynamics* (Prentice-Hall, Upper Saddle River, NJ, 1999)
69. J. Sticklen, M.T. Eskil, *An Introduction to Technical Problem Solving with MATLAB* (Great Lakes Press, Wildwood, MO, 2006)
70. A.C. Ugural, *Mechanical Design* (McGraw-Hill, New York, 2004)
71. R. Voinea, D. Voiculescu, V. Ceausu, *Mechanics* (Mecanica) (EDP, Bucharest, 1983)

72. K.J. Waldron, G.L. Kinzel, *Kinematics, Dynamics, and Design of Machinery* (Wiley, New York, 1999)
73. J.H. Williams Jr., *Fundamentals of Applied Dynamics* (Wiley, New York, 1996)
74. C.E. Wilson, J.P. Sadler, *Kinematics and Dynamics of Machinery* (Harper Collins College Publishers, New York, 1991)
75. H.B. Wilson, L.H. Turcotte, and D. Halpern, *Advanced Mathematics and Mechanics Applications Using MATLAB* (Chapman & Hall/CRC, 2003)
76. S. Wolfram, *Mathematica* (Wolfram Media/Cambridge University Press, Cambridge, 1999)
77. National Council of Examiners for Engineering and Surveying (NCEES), *Fundamentals of Engineering. Supplied-Reference Handbook* (Clemson, SC, 2001)
78. * * * , *The Theory of Mechanisms and Machines* (Teoria mehanizmov i masin), Vassaia scola, Minsk, Russia (1970)
79. eCourses - University of Oklahoma: http://ecourses.ou.edu/home.html
80. https://www.mathworks.com
81. http://www.eng.auburn.edu/~marghitu/
82. https://www.solidworks.com
83. https://www.wolfram.com

Chapter 7
Epicyclic Gear Trains

Abstract Main components of the gears are defined. The angular velocities of epicyclic gear trains are calculated using the classical method and the closed contour method.

7.1 Introduction

Gears are toothed elements that transmit rotational motion from one shaft to another. The law of conjugate gear-tooth action states the common normal to the surfaces at the point of contact of two gears in rotation must always intersect the line of centers at the same point P, called the pitch point, Fig. 7.1. The pitch circles are the circles that remain tangent throughout the gear motion. The point of tangency is the pitch point. The diameter of the pitch circle is the pitch diameter. The diameter of a gear always denotes its pitch diameter, d. The involute is defined as the locus of a point on a taut string that is unwrapped from a cylinder. The circle that represents the cylinder is the base circle. The involute profiles of gears are increased by a distance, a, called the addendum. The outer circle is the addendum circle with the radius $r_a = r + a$.

The circular pitch is designated as p, and is measured in inches (English units) or millimeters (SI units)

$$p = \pi d/N, \quad p = \pi d_p/N_p, \quad p = \pi d_g/N_g, \tag{7.1}$$

where N is the number of teeth in the gear (or pinion), N_p is the number of teeth in the pinion, and N_g is the number of teeth in the gear. The pitch diameter of the pinion is d_p and the pitch diameter of the gear is d_g.

The diametral pitch P_d (English units), is defined as the number of teeth per inch of pitch diameter

© The Author(s), under exclusive license to Springer Nature Switzerland AG 2022
D. B. Marghitu et al., *Mechanical Simulation with MATLAB®*,
Springer Tracts in Mechanical Engineering,
https://doi.org/10.1007/978-3-030-88102-3_7

165

Fig. 7.1 Geometry of a gear

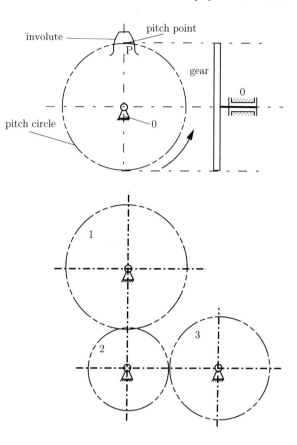

Fig. 7.2 Simple gear train

$$P_d = N/d, \quad P_d = N_p/d_p, \quad P_d = N_g/d_g. \tag{7.2}$$

The module m (used only with SI), is the complementarity of P_d, and is defined as the pitch diameter in millimeters divided by the number of teeth (number of millimeters of pitch diameter per tooth)

$$m = d/N, \quad m = d_p/N_p, \quad m = d_g/N_g. \tag{7.3}$$

A gear train is any collection of two or more meshing gears. Figure 7.2 shows a simple gear train with three gears in series. The train ratio is computed with the relation

$$i_{13} = \frac{\omega_1}{\omega_3} = \frac{\omega_1}{\omega_2}\frac{\omega_2}{\omega_3} = \left(-\frac{N_2}{N_1}\right)\left(-\frac{N_3}{N_2}\right) = \frac{N_3}{N_1}. \tag{7.4}$$

Only the sign of the overall ratio is affected by the intermediate gear 2 which is an idler.

When at least one of the gear axes rotates relative to a frame in addition to the gear's own rotation about its own axes, the train is called a planetary gear train or epicyclic gear train. When a generating circle (planet gear) rolls on the outside of another circle, called a directing circle (sun gear), each point on the generating circle describes an epicycloid.

7.2 Epicyclic Gear Train with One Planet

A planetary gear train is shown in the Fig. 7.3(a). The sun gear 1 has N1 teeth. The planet gear 2 has N2 teeth. The ring gear 4 is fixed. The module of the gears is m. The sun gear 1 has an angular speed of n1 rpm. The numerical values are:

```
N1 = 30; % no. of teeth of sun gear 1
N2 = 15; % no. of teeth of planet gear 2
```

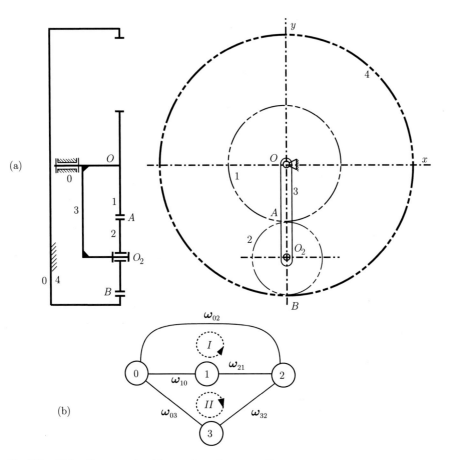

Fig. 7.3 a Epicyclic gear train with one planet; **b** contour diagram

```
m = 22*10^-3; % (m) the module of the gears

n1 = 50; % (rpm) angular velocity of 1
omega1 = n1*pi/30;    % (rad/s)
omega1_ = [0 0 omega1];
```

Determine the angular speed of the arm 3. The pitch circle radii of the gears and the number of teeth of the ring gear are:

```
% pitch radii
r1 = m*N1/2; % (m) pitch radius of sun gear 1
r2 = m*N2/2; % (m) pitch radius of planet gear 1
r4 = r1 + 2*r2;   % (m)  pitch radius of ring gear 4
N4 = 2*r4/m; % no. of teeth of ring gear 4

% r1 =  0.330 (m)
% r2 =  0.165 (m)
% r4 =  0.660 (m)
% N4 = 60 (teeth)
```

The position vectors of important joints are:

```
% origin of the reference frame
% position vector of center of gear 1: O1=O
rO1_ = [0 0 0];
% position vector of center of gear 4: O4
rO4_ = rO1_;
% position vector of center of planet gear 2: O2
rO2_ = [0 -r1-r2 0];
% position vector of pitch point A (gears 1 and 2)
rA_ = [0 -r1 0];
% position vector of pitch point B (gears 2 and 4)
rB_ = [0 -r4 0];

% rO1_=rO4_=rO_= [0 0 0] (m)
% rO2_ = [0 -0.495 0] (m)
% rA_ = [0 -0.330 0] (m)
% rB_ = [0 -0.660 0] (m)
```

7.2.1 Classical Method

The velocity of the pitch point A_1 on the sun gear 1 is:

```
vO_ = [0 0 0]; % velocity of O on sun gear 1
% velocity of A=A1 on sun gear 1
vA1_ = vO_ + cross(omega1_, rA_);
% vA1_ = [ 1.728 0 0] (m/s)
```

The velocity of the pitch point A_2 on the planet gear 2 is:

```
% omega2 unknown
syms omega2
omega2_ = [0 0 omega2]; % angular velocity of planet 2
% vB2_ is the velocity of B2 on planet gear 2
% vB4_ is the velocity of B4 on fixed ring gear 4
% vB2_ = vB4_ = [0 0 0]
vB2_ = [0 0 0];
% velocity of A2 on planet gear 2
vA2_ =  vB2_ + cross(omega2_, rA_-rB_);
```

The pitch line velocities at A are equal and the angular velocity of the planet gear is:

```
% vA1_ = vA2_
eqvA_ = vA1_-vA2_;
eqvA  = eqvA_(1); % (1)
% Eq.(1) =>
omega2  = eval(solve(eqvA));
omega2_ = [0 0 omega2];
% omega2=-5.236(rad/s)=-50.000(rpm)
```

The angular velocity of the planet arm 3 is calculated with:

```
% omega3 unknown
syms omega3
omega3_ = [0 0 omega3]; % angular velocity of planet arm 3
% vO22_ is the velocity of O2=O22 on planet gear 2
% vO23_ is the velocity of O2=O23 on planet arm 3
vO22_ = vB2_ + cross(omega2_,rO2_-rB_);
vO23_ = vO_ + cross(omega3_,rO2_);
% vO22_ = vO23_
eqvO2_ = vO22_-vO23_;
eqvO2 = eqvO2_(1); % (2)
% Eq.(2) =>
omega3  = eval(solve(eqvO2));
omega3_ = [0 0 omega3];
% omega3= 1.745(rad/s)=16.667(rpm)
```

7.2.2 Contour Method

The number of independent loops is

$$n_c = c - n = 5 - 3 = 2,$$

where n is the number of moving links (gear 1, gear 2, and arm 3) and c is the number of joints. Between the fixed frame 0 and the sun gear 1 there is a revolute joint at O, between 1 and the planet gear 2 there is a gear joint at A, between the arm 3 and 2 there is a revolute joint at O_2, between 2 and the fixed ring gear 4 there is a gear joint at B, and between 3 and 0 there is a revolute joint at O. There are two independent contours, as shown in Fig. 7.3(b).

The velocity equations for a simple closed kinematic chain are [3, 71]

$$\sum_{(i)} \boldsymbol{\omega}_{i,i-1} = \mathbf{0} \text{ and } \sum_{(i)} \mathbf{r}_{A_i} \times \boldsymbol{\omega}_{i,i-1} = \mathbf{0},$$

where $\boldsymbol{\omega}_{i,i-1}$ is the relative angular velocity of the rigid body (i) with respect to the rigid body $(i-1)$ and $\mathbf{r}_{A_i}$ is the position vector of the joint between the rigid body (i) and the rigid body $(i-1)$ with respect to a fixed reference frame.

For the first closed contour 0, 1, 2, and 4=0 (counterclockwise) the velocity vectorial equations are

$$\boldsymbol{\omega}_{10} + \boldsymbol{\omega}_{21} + \boldsymbol{\omega}_{02} = \mathbf{0},$$
$$\mathbf{r}_A \times \boldsymbol{\omega}_{21} + \mathbf{r}_B \times \boldsymbol{\omega}_{02} = \mathbf{0},$$

or

$$\boldsymbol{\omega}_{10} + \boldsymbol{\omega}_{21} + \boldsymbol{\omega}_{02} = 0,$$

$$\begin{vmatrix} \mathbf{\imath} & \mathbf{\jmath} & \mathbf{k} \\ 0 & -r_1 & 0 \\ 0 & 0 & \omega_{21} \end{vmatrix} + \begin{vmatrix} \mathbf{\imath} & \mathbf{\jmath} & \mathbf{k} \\ 0 & -r_4 & 0 \\ 0 & 0 & \omega_{02} \end{vmatrix} = \mathbf{0}.$$

The MATLAB program for the first contour is:

```
% contour 0-1-2-4=0
omega10_ = [0 0 omega1 ];
syms omega21 omega02
% omega21 is the relative angular velocity of 2 with respect to 1 at A
% omega02 is the relative angular velocity of 0=4 with respect to 2 at B
omega21_ = [0 0 omega21];
omega02_ = [0 0 omega02];
% omega10_ + omega21_ + omega02_ = 0_
% OA_ x omega21_ + OB_ x omega02_ = 0_
eqI1_ = omega10_ + omega21_ + omega02_;
eqI1 = eqI1_(3); % (3)
eqI2_ = cross(rA_,omega21_)+cross(rB_,omega02_);
```

```
eqI2 = eqI2_(1); % (4)

% Eqs.(3)(4)=>
solI = solve(eqI1,eqI2);
omega21 = eval(solI.omega21);
omega02 = eval(solI.omega02);
omega20 = - omega02;

% omega21 = -10.472 (rad/s)
% omega20 = -5.236 (rad/s)
```

The second closed contour is 0, 1, 2, 3 and 0 (clockwise path). The vectorial
equations are

$$\boldsymbol{\omega}_{10} + \boldsymbol{\omega}_{21} + \boldsymbol{\omega}_{32} + \boldsymbol{\omega}_{03} = \mathbf{0},$$
$$\mathbf{r}_A \times \boldsymbol{\omega}_{21} + \mathbf{r}_{O_2} \times \boldsymbol{\omega}_{32} = \mathbf{0},$$

or

$$\omega_{10} + \omega_{21} + \omega_{32} + \omega_{03} = 0,$$

$$\begin{vmatrix} \mathbf{i} & \mathbf{j} & \mathbf{k} \\ 0 & -r_1 & 0 \\ 0 & 0 & \omega_{21} \end{vmatrix} + \begin{vmatrix} \mathbf{i} & \mathbf{j} & \mathbf{k} \\ 0 & -r_1 - r_2 & 0 \\ 0 & 0 & \omega_{32} \end{vmatrix} = \mathbf{0}.$$

The MATLAB commands for the second contour are:

```
% contour 0-1-2-3-0
omega21_ = [0 0 omega21];
syms omega32 omega03
% omega32 is the relative angular velocity of 3 with respect to 2 at O2
% omega03 is the relative angular velocity of 0 with respect to 3 at O
omega32_ = [0 0 omega32];
omega03_ = [0 0 omega03];

% omega10_ + omega21_ + omega32_ + omega03_ = 0_
% OA_ x omega21_ + OO2_ x omega32_ = 0_

eqII1_ = omega10_ + omega21_ + omega32_ + omega03_;
eqII1 = eqII1_(3); % (5)
eqII2_ = cross(rA_, omega21_)+cross(rO2_, omega32_);
eqII2 = eqII2_(1); % (6)
% Eqs.(5)(6)=>
solII = solve(eqII1,eqII2);
omega32 = eval(solII.omega32);
omega03 = eval(solII.omega03);
omega30 = - omega03;

% omega32 =  6.981 (rad/s)
% omega30 =  1.745 (rad/s)
```

7.3 Mechanism with Epicyclic Gears

The driver link 1 (AB) has the angular velocity w1, as shown in Fig. 7.4a. The gears 2 and 3 are attached to the oscillatory link 4 ($FCDE$). The input data for the mechanism is:

```
AB = 0.050; % (m) AB is link 1
% gear with the center at C is link 2
% gear with the center at E is link 3
% FCDE is link 4
CF = 0.070; % (m) link 4
CD = 0.050; % (m) link 4
DE = 0.025; % (m) link 4
CB = 0.030; % (m)

% A=O origin
% coordinates of fixed joint F
xF = AB;
yF = -0.0632;
rF_ = [xF yF 0]; % position vector of F

w1 = 5; % (rad/s) angular velocity of link 1

theta = pi/6; % (rad) angle of 1 with x-axis
```

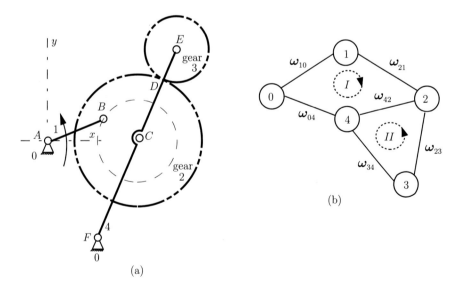

(a)

(b)

Fig. 7.4 a Mechanism with epicyclic gears; **b** contour diagram

The angular velocities of gears 2 and 3 will be calculated at this instant. The position of the revolute joint B between link 1 and gear 2 is:

```
xB = AB*cos(theta);
yB = AB*sin(theta);
rB_ = [xB yB 0]; % position vector of B
% rB_ =  [  0.043,   0.025, 0 ] (m)
```

The position of joint C, the center of gear 2, is calculated with:

```
% joint C
syms xCsol yCsol
eqnC1 = (xCsol - xB )^2 + ( yCsol - yB )^2 - CB^2;
eqnC2 = (xCsol - xF )^2 + ( yCsol - yF )^2 - CF^2;
solC_ = solve(eqnC1, eqnC2, xCsol, yCsol);
xCpositions = eval(solC_.xCsol);
yCpositions = eval(solC_.yCsol);
xC1 = xCpositions(1);
xC2 = xCpositions(2);
yC1 = yCpositions(1);
yC2 = yCpositions(2);
% select the correct position for C
% for the given input angle theta
if xC1 > xF
    xC = xC1; yC=yC1;
else
    xC = xC2; yC=yC2;
end
rC_ = [xC yC 0]; % position vector of C
% rC_ =  [  0.066,   0.005, 0 ] (m)
```

The angle of link 4 with the horizontal is:

```
% angle of link 4 with x-axis
phi = atan((yF-yC)/(xF-xC));
% phi = 77.055 (degrees)
```

The position vectors of D, the pitch point of the gears, and E, the revolute joint of the center of gear 3, are:

```
xD = xC+CD*cos(phi);
yD = yC+CD*sin(phi);
rD_ = [xD yD 0]; % position vector of C

xE = xD+DE*cos(phi);
```

```
yE = yD+DE*sin(phi);
rE_ = [xE yE 0]; % position vector of C

% rD_ = [  0.077,   0.054, 0 ] (m)
% rE_ = [  0.082,   0.078, 0 ] (m)
```

7.3.1 Classical Method—Velocity Analysis

The velocity of *B* is calculated with:

```
w1_ = [0 0 w1]; % angular velocity of link 1
vA_ = [0 0  0]; % velocity of A (fixed)
% velocity of B (A and B on the same body 1)
vB_ = vA_ + cross(w1_, rB_);
% vB_ = [ -0.125,   0.217, 0 ] (m/s)
```

The angular velocities of gear 2 and link 4 are calculated with:

```
syms omega2 omega4
omega2_ = [0 0 omega2]; % angular velocity of gear 2
omega4_ = [0 0 omega4]; % angular velocity of link 4
vF_ = [0 0 0]; % velocity of F (fixed)

% vC2_ is the velocity of C on gear 2 (C=C2)
vC2_ =  vB_ + cross(omega2_, rC_-rB_);
% vC4_ is the velocity of C on link 4 (C=C4)
vC4_ =  vF_ + cross(omega4_, rC_-rF_);
% C is a revolute joint => vC2_ = vC4_
eqvC_ = vC2_ - vC4_;
eqvCx  = eqvC_(1); % (1)
eqvCy  = eqvC_(2); % (2)
% Eqs.(1)(2) =>
solvC = solve(eqvCx, eqvCy);
omega2  = eval(solvC.omega2);
omega4  = eval(solvC.omega4);
fprintf('omega2 = %6.3f (rad/s)\n',omega2);
fprintf('omega4 = %6.3f (rad/s)\n',omega4);
omega2_ = [0 0 omega2];
omega4_ = [0 0 omega4];

% omega2 = -6.962 (rad/s)
% omega4 =  3.871 (rad/s)
```

The angular velocity of gear 3 is computed with:

```
syms omega3
omega3_ = [0 0 omega3];  % angular velocity of gear 3
% vE4_ is the velocity of E on link 4  (C=C4)
vE4_ = vF_  + cross(omega4_, rE_-rF_);
% E is a revolute joint => vE3_ = vE4_
vE3_ = vE4_;
% vD3_ is the velocity of D on gear 3  (D=D3)
vD3_ = vE3_ + cross(omega3_, rD_-rE_);
% vC2_ is the velocity of C on gear 2  (C=C2)
vC2_ = vB_  + cross(omega2_, rC_-rB_);
% vD2_ is the velocity of D on gear 2  (D=D2)
vD2_ = vC2_ + cross(omega2_, rD_-rC_);
% vD2_ = vD3_
eqvD_ = vD2_ - vD3_;
eqvDx  = eqvD_(1);  % (3)
eqvDy  = eqvD_(2);
% Eq.(3)=>
omega3 = eval(solve(eqvDx));
fprintf('omega3 = %6.3f (rad/s)\n',omega3);
% omega3 = 25.537 (rad/s)
```

7.3.2 Contour Method—Velocity Analysis

The same results are obtained using the contour method. The first contour is
0->1->2->4->0 , as shown in Fig. 7.4b, clockwise path. The unknowns are the
relative angular velocities:

```
syms w21 w42 w04
w10_ = [0, 0, w1 ];
% w21 is the relative angular velocity of 2 with respect to 1 at B
w21_ = [0, 0, w21];
% w42 is the relative angular velocity of 4 with respect to 2 at C
w42_ = [0, 0, w42];
% w04 is the relative angular velocity of 0 with respect to 4 at F
w04_ = [0, 0, w04];
```

The velocity contour equations are:

```
eq11_ = w10_ + w21_ + w42_ + w04_;
eq11z = eq11_(3);  % (4)
eq12_ = cross(rB_, w21_) + cross(rC_, w42_) + cross(rF_, w04_);
eq12x = eq12_(1);  % (5)
```

```
eq12y = eq12_(2); % (6)
```

and the results are:

```
% Eqs.(4)(5)(6)=>
sol1 = solve(eq11z, eq12x, eq12y);
w21 = eval(sol1.w21);
w42 = eval(sol1.w42);
w04 = eval(sol1.w04);

w40 = -w04;      % angular velocity of link 4
w20 = w1 + w21; % angular velocity of gear 2

% w21 = -11.962 (rad/s)
% w42 =  10.833 (rad/s)
% w40 =   3.871 (rad/s)
% w20 =  -6.962 (rad/s)
```

The angular velocity of gear 3 is calculated from the second contour 2->4->3->2, as shown in Fig. 7.4(b), counterclockwise path:

```
syms w34 w23
w42_ = [0, 0, w42];
% w34 is the relative angular velocity of 3 with respect to 4 at E
w34_ = [0, 0, w34];
% w23 is the relative angular velocity of 2 with respect to 3 at D
w23_ = [0, 0, w23];

eq21_ = w42_ + w34_ + w23_;
eq21z = eq21_(3); % (7)
eq22_ = cross(rC_, w42_) + cross(rE_, w34_) + cross(rD_, w23_);
eq22x = eq22_(1); % (8)
eq22y = eq22_(2); % (9)

% Eqs.(7)(8)=>
sol2 = solve(eq21z, eq22x);
w34 = eval(sol2.w34);
w23 = eval(sol2.w23);

w30 = w40 + w34;
w3 = w20 - w23;  % angular velocity of gear 3

fprintf('w23 = %6.3f (rad/s)\n',w23)
fprintf('w34 = %6.3f (rad/s)\n',w34)
fprintf('w30 = %6.3f (rad/s)\n',w30)

% w23 = -32.499 (rad/s)
% w34 =  21.666 (rad/s)
% w30 =  25.537 (rad/s)
```

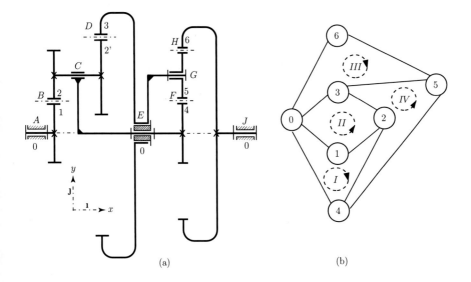

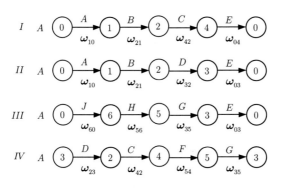

Fig. 7.5 a Epicyclic gear train with multiple planets; **b** contour diagram

7.4 Epicyclic Gear Train with Multiple Planets

A planetary gear train is shown in Fig. 7.5a. The input data is:

```
m = 24; % (mm) module of the gears
m = m*10^-3; % (m)

N1 = 22; % (teeth) no. of teeth of sun gear 1
N2 = 20; % (teeth) no. of teeth of planet gear 2
N2p = 35;% (teeth) no. of teeth of planet gear 2'
N4 = 15; % (teeth) no. of teeth of sun gear 4
```

```
N5 = 16; % (teeth) no. of teeth of planet gear 5

n3 = 200; % (rpm) angular velocity of ring gear 3
n6 = 150; % (rpm) angular velocity of ring gear 6

w3 = pi*n3/30; % (rad/s)
w6 = pi*n6/30; % (rad/s)

% angular velocities are along x-axis
omega3_ = [w3, 0, 0]; % angular velocity vector of 3
omega6_ = [w6, 0, 0]; % angular velocity vector of 6
```

The pitch radii of the gears and the number of teeth for the ring gears are:

```
r1 = m*N1/2; % (m) pitch radius of sun gear 1
r2 = m*N2/2; % (m) pitch radius of planet gear 2
r2p = m*N2p/2; % (m) pitch radius of planet gear 2'
r3 = r1 + r2 + r2p;  % (m) pitch radius of ring gear 3
r4 = m*N4/2; % (m) pitch radius of planet gear 4
r5 = m*N5/2; % (m) pitch radius of planet gear 5
r6 = r4 + 2*r5; % (m) pitch radius of ring gear 6

N3 = 2*r3/m; % (teeth) no. of teeth of ring gear 3
N6 = 2*r6/m; % (teeth) no. of teeth of ring gear 6

% r1 =   0.264 (m)
% r2 =   0.240 (m)
% r2p =  0.420 (m)
% r3 =   0.924 (m)
% r4 =   0.180 (m)
% r5 =   0.192 (m)
% r6 =   0.564 (m)
% N3 = 77 (teeth)
% N6 = 47 (teeth)
```

The position vectors of the joints are:

```
% position vectors
syms xC xD xE xF xG xH xJ
% origin of the reference frame
% position vector of center of gear 1
rA_ = [0, 0, 0];
% position vector of pitch point B (gears 1 and 2)
rB_ = [0, r1, 0];
% position vector of joint C
```

```
rC_ = [xC, r1 + r2, 0];
% position vector of pitch point D (gears 2' and 3)
rD_ = [xD, r1 + r2 + r2p, 0];
% position vector of joint E
rE_ = [xE, 0, 0];
% position vector of pitch point F (gears 4 and 5)
rF_ = [xF, r4, 0];
% position vector of joint G
rG_ = [xG, r4 + r5, 0];
% position vector of pitch point H (gears 5 and 6)
rH_ = [xH, r4 + 2*r5, 0];
% position vector of joint J
rJ_ = [xJ, 0, 0];
```

The velocities of the fixed joints are:

```
vA_ = [0 0 0]; % velocity vector of A
vE_ = [0 0 0]; % velocity vector of E
vJ_ = [0 0 0]; % velocity vector of J
```

7.4.1 *Classical Method—Velocity Analysis*

The unknown angular velocities are:

```
syms w1 w2 w4 w5
omega1_ = [w1, 0, 0]; % angular velocity of 1
omega2_ = [w2, 0, 0]; % angular velocity of 2
omega4_ = [w4, 0, 0]; % angular velocity of 4
omega5_ = [w5, 0, 0]; % angular velocity of 5
```

The angular velocity of the planet gear 5 is calculated with:

```
% G and E are on link 3
vG_  = vE_ + cross(omega3_, rG_-rE_);
% H and J are on link 6
vH6_ = vJ_ + cross(omega6_, rH_-rJ_); % link 6
vH5_ = vH6_;
% G and H are on link 5
vG5_ = vH5_ + cross(omega5_, rG_-rH_); % link 5
% velocity of G on link 5 is equal to velocity of G on link 3
eqw5_= vG5_-vG_; % (1)
% Eq.(1) =>
w5 = eval(solve(eqw5_(3)));
omega5_ = [w5 0 0];
% w5 =   5.563 (rad/s)
```

The angular velocity of the sun gear 4 is determined with:

```
% F and G are on link 5
vF_  = vG_ + cross(omega5_, rF_-rG_); % link 5
% F and E are on link 4
vF4_ = vE_ + cross(omega4_, rF_-rE_); % link 4
% velocity of F on link 5 is equal to velocity of F on link 4
eqw4_= vF4_-vF_; % (2)
% Eq.(2) =>
w4 = eval(solve(eqw4_(3)));
omega4_ = [w4 0 0];
% w4 = 37.350 (rad/s)
```

The angular velocity of the planet gears 2 and 2' is resolved with:

```
% D and E are on link 3
vD_ = vE_ + cross(omega3_, rD_-rE_); % link 3
% C and E are on link 4
vC_ = vE_ + cross(omega4_, rC_-rE_); % link 4
% D and C are on link 2
vD2_= vC_ + cross(omega2_, rD_-rC_); % link 2
% velocity of D on link 2 is equal to velocity of D on link 3
eqw2_= vD2_ - vD_; % (3)
% Eq.(3) =>
w2 = eval(solve(eqw2_(3)));
omega2_ = [w2 0 0];
% w2 =  1.257 (rad/s)
```

The sun gear 1 has an angular velocity calculated with:

```
% B and D are on link 2
vB_ = vD_ + cross(omega2_, rB_-rD_); % link 2
% B and A are on link 1
vB1_= vA_ + cross(omega1_, rB_-rA_); % link 1
% velocity of B on link 1 is equal to velocity of B on link 2
eqw1_= vB1_ - vB_; % (4)
% Eq.(4) =>
w1 = eval(solve(eqw1_(3)));
omega1_ = [w1 0 0];
% w1 = 70.162 (rad/s)
```

7.4.2 Contour Method—Velocity Analysis

There are four independent contours for the gear train, as shown in Fig. 7.5b. The given relative angular velocity are:

```
w03 = -w3; % relative angular velocity of 0 with respect to 3
w60 = w6;  % relative angular velocity of 6 with respect to 0
```

For the first contour $0->1->2->4->0$, as shown in Fig. 7.5b, clockwise path, the following equations can be written:

```
% contour I: 0-A-1-B-2-C-4-E-0 cw
syms w10 w21 w42 w04
omega10_ = [w10, 0, 0]; % relative angular velocity of 1 with respect to 0
omega21_ = [w21, 0, 0]; % relative angular velocity of 2 with respect to 1
omega42_ = [w42, 0, 0]; % relative angular velocity of 4 with respect to 2
omega04_ = [w04, 0, 0]; % relative angular velocity of 0 with respect to 4

% omega10_+omega21_+omega42_+omega04_=0_
eq11_ = omega10_ + omega21_ + omega42_ + omega04_;
eq1x = eq11_(1); % (5)
% rB_ x omega21_ + rC_ x omega42_ = 0_
eq12_ = cross(rB_, omega21_) + cross(rC_, omega42_);
eq1z = eq12_(3); % (6)
```

The relative velocity equations for the second contour $0->1->2->3->0$, counterclockwise path, are:

```
% contour II: 0-A-1-B-2-D-3-E-0 ccw
syms w32
omega32_ = [w32, 0, 0]; % relative angular velocity of 3 with respect to 2
omega03_ = [w03, 0, 0]; % relative angular velocity of 0 with respect to 3

% omega10_+omega21_+omega32+omega03=0_
eq21_ = omega10_ + omega21_ + omega32_ + omega03_;
eq2x = eq21_(1);  % (7)
% rB_ x omega21_ + rD_ x omega32_ = 0_
```

```
eq22_ = cross(rB_, omega21_) + cross(rD_, omega32_);
eq2z = eq22_(3);   % (8)
```

For the third contour `0->6->5->3->0`, clockwise path, the relative angular velocity equations are:

```
% contour III: 0-J-6-H-5-G-3-E-0 cw
syms   w56 w35
omega60_ = [w60, 0, 0]; % relative angular velocity of 6 with respect to 0
omega56_ = [w56, 0, 0]; % relative angular velocity of 5 with respect to 6
omega35_ = [w35, 0, 0]; % relative angular velocity of 3 with respect to 5

% omega60_+omega56_+omega35_+omega03_=0_
eq31_ = omega60_ + omega56_ + omega35_ + omega03_;
eq3x = eq31_(1);   % (9)
% rH_ x omega56_ + rG_ x omega35_ = 0_
eq32_ = cross(rH_, omega56_) + cross(rG_, omega35_);
eq3z = eq32_(3);   % (10)
```

For the last contour `3->2->4->5->3`, counterclockwise path, the relative angular velocity equations are:

```
% contour IIII: 3-D-2-C-4-F-5-G-3 ccw
syms w54
omega23_ = -omega32_; % relative angular velocity of 2 with respect to 3
omega54_ = [w54, 0, 0]; % relative angular velocity of 5 with respect to 4

% omega23_+omega42_+omega54_+omega35_=0_
eq41_ = omega23_ + omega42_ + omega54_ + omega35_;
eq4x = eq41_(1);    % (11)
% rD_ x omega23_ + rC_ x omega42_ + rF_ x omega54_ + rG_ x omega35_ = 0_
eq42_ = cross(rD_, omega23_) + cross(rC_, omega42_) ...
    +cross(rF_, omega54_) + cross(rG_, omega35_);
eq4z = eq42_(3);   % (12)
```

Solving all the closed contour equations, the angular velocities are obtained:

```
% Eqs.(5)(6)(7)(8)(9)(10)(11)(12)=>
sys= [eq1x eq1z eq2x eq2z eq3x eq3z eq4x eq4z];
sol = solve(sys);

w10 = eval(sol.w10);
w21 = eval(sol.w21);
w42 = eval(sol.w42);
w04 = eval(sol.w04);
w32 = eval(sol.w32);
w35 = eval(sol.w35);
w56 = eval(sol.w56);
w54 = eval(sol.w54);


w2 = w10+w21; % angular velocity of 2
```

```
w4 = -w04;     % angular velocity of 4
w5 = w4+w54;   % angular velocity of 5

% w1 = 70.162 (rad/s)
% w2 =  1.257 (rad/s)
% w4 = 37.350 (rad/s)
% w5 =  5.563 (rad/s)
%
% n1 = 670.000 (rpm)
% n2 = 12.000 (rpm)
% n4 = 356.667 (rpm)
% n5 = 53.125 (rpm)
```

7.5 Problems

7.1 A planetary gear train is shown in Fig. 7.6. The sun gear 1 has $N_1 = 16$ teeth. The planet gear 2 has $N_2 = 30$ teeth. The gear 4 has $N_4 = 22$ teeth. The ring gear 3' has $N_{3'} = 84$ teeth. The module of the gears is $m = 22$ mm. Gear 1 rotates with an input angular speed $n_1 = 120$ rpm. Find the angular speed of the gear 4, n_4 using the presented two methods.

Fig. 7.6 Problem 7.1

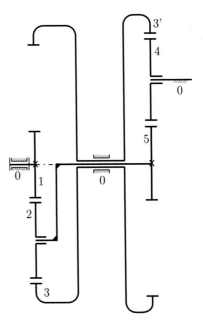

Fig. 7.7 Problem 7.2

Fig. 7.8 Problem 7.3

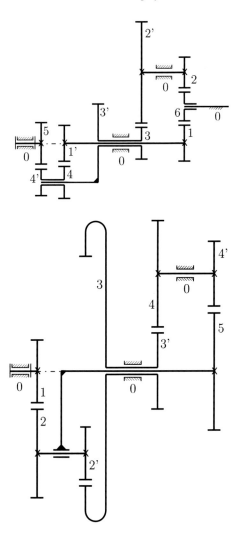

7.2 A planetary gear train is shown in Fig. 7.7. The gear 1 has $N_1 = 24$ teeth. The
    gear 2 has $N_2 = 22$ teeth. The gear 3 has $N_3 = 20$ teeth. The gear 1' has $N_{1'} = 32$
    teeth. The planet gear 4 has $N_4 = 28$ teeth. The gear 5 has $N_5 = 24$ teeth. The
    gear 6 has $N_6 = 19$ teeth. The module of the gears is $m = 10$ mm. Gear 1 has
    an input angular speed $n_1 = 180$ rpm. Find the angular speed of the gear 5, n_5.
    Compare the results obtained with the classic method and the results from the
    contour method.

7.3 A planetary gear train is shown in Fig. 7.8. The sun gear 1 has $N_1 = 18$ teeth.
    The planet gear 2 has $N_2 = 42$ teeth. The planet gear 2' has $N_{2'} = 21$ teeth.
    The gear 3' has $N_{3'} = 21$ teeth. The gear 4 has $N_4 = 28$ teeth. The gear 4' has
    $N_{4'} = 14$ teeth. The module of the gears is $m = 24$ mm. Gear 1 rotates with an

Fig. 7.9 Problem 7.4

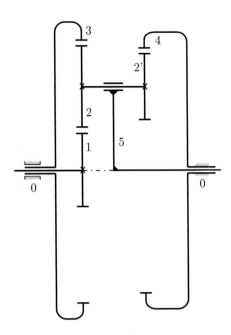

input angular speed $n_1 = -160$ rpm. Find the angular speed of the gear 5, n_5 using the classical and contour methods.

7.4 A planetary gear train is shown in Fig. 7.9. The sun gear 1 has $N_1 = 20$ teeth. The planet gear 2 has $N_2 = 29$ teeth. The planet gear 2' has $N_{2'} = 21$ teeth. The module of the gears is $m = 20$ mm. The sun gear 1 has an angular speed of

Fig. 7.10 Problem 7.5

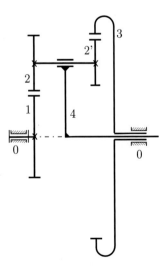

$n_1 = 273$ rpm. Find the angular speeds of the gear 3, n_3, and the gear 4, n_4. Use the classical and contour methods.

7.5 A planetary gear train is shown in Fig. 7.10. The sun gear 1 has $N_1 = 20$ teeth. The planet gear 2 has $N_2 = 13$ teeth. The planet gear 2' has $N_{2'} = 11$ teeth. Gears 2 and 2' are fixed on the same shaft. The module of the gears is $m = 30$ mm. Gear 1 rotates with an input angular speed $n_1 = 200$ rpm. Gear 3 rotates at $n_3 = -160$ rpm. Using the classical and contour methods find the angular speeds of the gear 2, n_2, and the bar 4, n_4.

References

1. E.A. Avallone, T. Baumeister, A. Sadegh, *Marks' Standard Handbook for Mechanical Engineers*, 11th edn. (McGraw-Hill Education, New York, 2007)
2. I.I. Artobolevski, *Mechanisms in Modern Engineering Design* (MIR, Moscow, 1977)
3. M. Atanasiu, *Mechanics* (Mecanica), EDP, Bucharest, 1973)
4. H. Baruh, *Analytical Dynamics* (WCB/McGraw-Hill, Boston, 1999)
5. A. Bedford, W. Fowler, *Dynamics* (Addison Wesley, Menlo Park, CA, 1999)
6. A. Bedford, W. Fowler, *Statics* (Addison Wesley, Menlo Park, CA, 1999)
7. F.P. Beer et al., *Vector Mechanics for Engineers: Statics and Dynamics* (McGraw-Hill, New York, NY, 2016)
8. M. Buculei, D. Bagnaru, G. Nanu, D.B. Marghitu, *Computing Methods in the Analysis of the Mechanisms with Bars* (Scrisul Romanesc, Craiova, 1986)
9. M.I. Buculei, *Mechanisms* (University of Craiova Press, Craiova, Romania, 1976)
10. D. Bolcu, S. Rizescu, *Mecanica* (EDP, Bucharest, 2001)
11. J. Billingsley, *Essential of Dynamics and Vibration* (Springer, 2018)
12. R. Budynas, K.J. Nisbett, *Shigley's Mechanical Engineering Design*, 9th edn. (McGraw-Hill, New York, 2013)
13. J.A. Collins, H.R. Busby, G.H. Staab, *Mechanical Design of Machine Elements and Machines*, 2nd Edn. (Wiley, 2009)
14. M. Crespo da Silva, *Intermediate Dynamics for Engineers* (McGraw-Hill, New York, 2004)
15. A.G. Erdman, G.N. Sandor, *Mechanisms Design* (Prentice-Hall, Upper Saddle River, NJ, 1984)
16. M. Dupac, D.B. Marghitu, *Engineering Applications: Analytical and Numerical Calculation with MATLAB* (Wiley, Hoboken, NJ, 2021)
17. A. Ertas, J.C. Jones, *The Engineering Design Process* (Wiley, New York, 1996)
18. F. Freudenstein, An application of boolean algebra to the motion of epicyclic drivers. J. Eng. Ind. 176–182 (1971)
19. J.H. Ginsberg, *Advanced Engineering Dynamics* (Cambridge University Press, Cambridge, 1995)
20. H. Goldstein, *Classical Mechanics* (Addison-Wesley, Redwood City, CA, 1989)
21. D.T. Greenwood, *Principles of Dynamics* (Prentice-Hall, Englewood Cliffs, NJ, 1998)
22. A.S. Hall, A.R. Holowenko, H.G. Laughlin, *Schaum's Outline of Machine Design* (McGraw-Hill, New York, 2013)
23. B.G. Hamrock, B. Jacobson, S.R. Schmid, *Fundamentals of Machine Elements* (McGraw-Hill, New York, 1999)
24. R.C. Hibbeler, *Engineering Mechanics: Dynamics*, Prentice Hall, 2010
25. T.E. Honein, O.M. O'Reilly, On the Gibbs–Appell equations for the dynamics of rigid bodies. J. Appl. Mech. **88**/074501-1 (2021)
26. R.C. Juvinall, K.M. Marshek, *Fundamentals of Machine Component Design*, 5th edn. (Wiley, New York, 2010)

27. T.R. Kane, *Analytical Elements of Mechanics*, vol. 1 (Academic, New York, 1959)
28. T.R. Kane, *Analytical Elements of Mechanics*, vol. 2 (Academic, New York, 1961)
29. T.R. Kane, D.A. Levinson, The use of kane's dynamical equations in robotics. MIT Int. J. Robot. Res. **3**, 3–21 (1983)
30. T.R. Kane, P.W. Likins, D.A. Levinson, *Spacecraft Dynamics* (McGraw-Hill, New York, 1983)
31. T.R. Kane, D.A. Levinson, *Dynamics* (McGraw-Hill, New York, 1985)
32. K. Lingaiah, *Machine Design Databook*, 2nd edn. (McGraw-Hill Education, New York, 2003)
33. N.I. Manolescu, F. Kovacs, A. Oranescu, *The Theory of Mechanisms and Machines* (EDP, Bucharest, 1972)
34. D.B. Marghitu, *Mechanical Engineer's Handbook* (Academic, San Diego, CA, 2001)
35. D.B. Marghitu, M.J. Crocker, *Analytical Elements of Mechanisms* (Cambridge University Press, Cambridge, 2001)
36. D.B. Marghitu, *Kinematic Chains and Machine Component Design* (Elsevier, Amsterdam, 2005)
37. D.B. Marghitu, *Mechanisms and Robots Analysis with MATLAB* (Springer, New York, N.Y., 2009)
38. D.B. Marghitu, M. Dupac, *Advanced Dynamics: Analytical and Numerical Calculations with MATLAB* (Springer, New York, N.Y., 2012)
39. D.B. Marghitu, M. Dupac, H.M. Nels, *Statics with MATLAB* (Springer, New York, N.Y., 2013)
40. D.B. Marghitu, D. Cojocaru, *Advances in Robot Design and Intelligent Control* (Springer International Publishing, Cham, Switzerland, 2016), pp. 317–325
41. D.J. McGill, W.W. King, *Engineering Mechanics: Statics and an Introduction to Dynamics* (PWS Publishing Company, Boston, 1995)
42. J.L. Meriam, L.G. Kraige, *Engineering Mechanics: Dynamics* (Wiley, New York, 2007)
43. C.R. Mischke, Prediction of Stochastic Endurance Strength. Transaction of ASME, Journal Vibration, Acoustics, Stress, and Reliability in Design **109**(1), 113–122 (1987)
44. L. Meirovitch, *Methods of Analytical Dynamics* (Dover, 2003)
45. R.L. Mott, *Machine Elements in Mechanical Design* (Prentice Hall, Upper Saddle River, NJ, 1999)
46. W.A. Nash, *Strength of Materials* Schaum's Outline Series. (McGraw-Hill, New York, 1972)
47. R.L. Norton, *Machine Design* (Prentice-Hall, Upper Saddle River, NJ, 1996)
48. R.L. Norton, *Design of Machinery* (McGraw-Hill, New York, 1999)
49. O.M. O'Reilly, *Intermediate Dynamics for Engineers Newton-Euler and Lagrangian Mechanics* (Cambridge University Press, UK, 2020)
50. O.M. O'Reilly, *Engineering Dynamics: A Primer* (Springer, NY, 2010)
51. W.C. Orthwein, *Machine Component Design* (West Publishing Company, St. Paul, 1990)
52. L.A. Pars, *A Treatise on Analytical Dynamics* (Wiley, New York, 1965)
53. F. Reuleaux, *The Kinematics of Machinery* (Dover, New York, 1963)
54. D. Planchard, M. Planchard, *SolidWorks 2013 Tutorial with Video Instruction* (SDC Publications, 2013)
55. I. Popescu, *Mechanisms* (University of Craiova Press, Craiova, Romania, 1990)
56. I. Popescu, C. Ungureanu, *Structural Synthesis and Kinematics of Mechanisms with Bars* (Universitaria Press, Craiova, Romania, 2000)
57. I. Popescu, L. Luca, M. Cherciu, D.B. Marghitu, *Mechanisms for Generating Mathematical Curves* (Springer Nature, Switzerland, 2020)
58. C.A. Rubin, *The Student Edition of Working Model* (Addison-Wesley Publishing Company, Reading, MA, 1995)
59. J. Ragan, D.B. Marghitu, Impact of a kinematic link with MATLAB and solidworks. Appl. Mech. Mater. **430**, 170–177 (2013)
60. J. Ragan, D.B. Marghitu, MATLAB Dynamics of a Free Link with Elastic Impact, in *International Conference on Mechanical Engineering, ICOME 2013*, May 16–17, 2013, Craiova, Romania (2013)
61. J.C. Samin, P. Fisette, *Symbolic Modeling of Multibody Systems* (Kluwer, 2003)
62. A.A. Shabana, *Computational Dynamics* (Wiley, New York, 2010)

63. I.H. Shames, *Engineering Mechanics - Statics and Dynamics* (Prentice-Hall, Upper Saddle River, NJ, 1997)
64. J.E. Shigley, C.R. Mischke, *Mechanical Engineering Design* (McGraw-Hill, New York, 1989)
65. J.E. Shigley, C.R. Mischke, R.G. Budynas, *Mechanical Engineering Design*, 7th edn. (McGraw-Hill, New York, 2004)
66. J.E. Shigley, J.J. Uicker, *Theory of Machines and Mechanisms* (McGraw-Hill, New York, 1995)
67. D. Smith, *Engineering Computation with MATLAB* (Pearson Education, Upper Saddle River, NJ, 2008)
68. R.W. Soutas-Little, D.J. Inman, *Engineering Mechanics: Statics and Dynamics* (Prentice-Hall, Upper Saddle River, NJ, 1999)
69. J. Sticklen, M.T. Eskil, *An Introduction to Technical Problem Solving with MATLAB* (Great Lakes Press, Wildwood, MO, 2006)
70. A.C. Ugural, *Mechanical Design* (McGraw-Hill, New York, 2004)
71. R. Voinea, D. Voiculescu, V. Ceausu, *Mechanics* (Mecanica), EDP, Bucharest, 1983)
72. K.J. Waldron, G.L. Kinzel, *Kinematics, Dynamics, and Design of Machinery* (Wiley, New York, 1999)
73. J.H. Williams Jr., *Fundamentals of Applied Dynamics* (Wiley, New York, 1996)
74. C.E. Wilson, J.P. Sadler, *Kinematics and Dynamics of Machinery* (Harper Collins College Publishers, New York, 1991)
75. H.B. Wilson, L.H. Turcotte, D. Halpern, *Advanced Mathematics and Mechanics Applications Using MATLAB* (Chapman & Hall/CRC, 2003))
76. S. Wolfram, *Mathematica* (Wolfram Media/Cambridge University Press, Cambridge, 1999)
77. National Council of Examiners for Engineering and Surveying (NCEES), *Fundamentals of Engineering. Supplied-Reference Handbook*, Clemson, SC (2001)
78. * * * , *The Theory of Mechanisms and Machines* (Teoria mehanizmov i masin), Vassaia scola, Minsk, Russia (1970)
79. eCourses - University of Oklahoma: http://ecourses.ou.edu/home.htm
80. https://www.mathworks.com
81. http://www.eng.auburn.edu/marghitu/
82. https://www.solidworks.com
83. https://www.wolfram.com

Chapter 8
Cam and Follower Mechanism

Abstract A cam and follower mechanism is analyzed. Same kinematic results are obtained with an equivalent mechanism. The differential method is used for velocity analysis.

A planar cam mechanism considered is shown in Fig. 8.1a. The driver link is the cam 1. The disk (or plate) cam 1 has a contour to make a specific motion to the follower 2. The follower moves in a plane perpendicular to the axis of rotation of the camshaft. The contact end of the follower 2 has a semicircular area with the radius R_f. The semicircle can be substituted with a roller. The follower maintains in contact with the cam. The center of the semicircle, C, must follow the profile of the cam 1. There are two instantaneous coincident points at C: the point C_2 is on the follower 2 and the point C_1 is on the cam path. The dashed line specifies the design profile of the cam. The contact point between the cam and the follower is P. The length of the follower is $L_f + R_f$, where $CP = R_f$, as shown in Fig. 8.1a.

The profile of the cam is a circle with the radius R_c and the center at $O = O_C$. The cam has a constant radius of curvature R_c. The cam is connected to the ground with a revolute joint at A. The distance between the center of the circle and the center of rotation is $AO = R_O$. The follower 2 has a vertical translational motion and is located at the distance e from the fixed point A from the vertical axis. The cam mechanism is an off-set cam-follower. If $e = 0$ the cam mechanism is a radial cam-follower.

The angle of AO with the horizontal axis is ϕ and the angle of $OC = R$ with the horizontal axis is ψ. The constant angular speed of the driver link 1 is n rpm.

The following numerical data are given: $OP = R_c = 0.1$ m, $e = 0.04$ m, $AO = R_O = 0.5 R_c$, $R_f = 0.0025$ m, and $n = 30$ rpm. The length L_f of the follower is $L_f = 0.15$ m and the width of the follower is $w = 2 R_f$. The depth of the cam and follower is $d = 0.01$ m. The material density of the cam and follower is $\rho = 8000$ kg/m^3.

© The Author(s), under exclusive license to Springer Nature Switzerland AG 2022 189
D. B. Marghitu et al., *Mechanical Simulation with MATLAB®*,
Springer Tracts in Mechanical Engineering,
https://doi.org/10.1007/978-3-030-88102-3_8

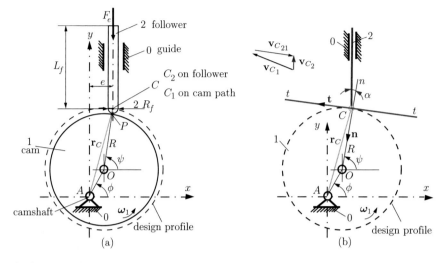

Fig. 8.1 a Cam and follower **b** sketch for kinematic analysis

The analysis is developed for the time when $\phi = \pi/3 = 60°$. At this time an external force of magnitude $F_e = 100$ N is acting on the follower as shown in Fig. 8.1a. The gravitational acceleration is $g = 9.807$ m/s^2.

At the point C there is a normal unit vector $\mathbf{n}$ and a tangent unit vector $\mathbf{t}$ at the profile of the cam, as shown in Fig. 8.1b. The angle between the normal to the profile of the cam and the follower is the pressure angle α. It is recommended for the pressure angle $\alpha < \pi/3$.

Next the kinematic and force analysis will be calculated.

8.1 Kinematics Analysis

The origin of the reference frame xy is selected at the joint A. The position of the center O of the cam 1 is calculated with

$$x_O = AO \cos \phi = R_O \cos \phi \quad \text{and} \quad y_O = AO \sin \phi = R_O \sin \phi.$$

The angle ψ of the segment OC with the horizontal is

$$\psi = \arccos \frac{e - x_O}{R},$$

where $R = R_c + R_f$. The position of the point C is given by

$$x_C = e \quad \text{and} \quad y_C = y_O + R \sin \psi \quad \text{and} \quad \mathbf{r}_C = x_C \mathbf{I} + y_C \mathbf{J}.$$

The MATLAB commands for the position of the mechanism are:

```
% revolute joint at A = O1
Rc = 0.1;      % (m) radius of the circle of the cam
e  = 0.04;     % (m) distance from follower to A
RO = 0.5*Rc;   % AO = RO
Rf = 0.0025;   % (m)
Lf = 0.15;     % (m) length of the follower
w  = 2*Rf;     % width of the follower
d  = 0.01;     % (m) depth of the cam and follower
rho= 8000;     % (kg/m^3) density of the material
g  = 9.807;    % (m/s^2) gravitational acceleration
Fe = 100;      % (N) external force on follower

% radius OC
R  = Rc + Rf;
phi = pi/3;
% position of O = Oc (rO_ = AO_)
xO = RO*cos(phi);  yO = RO*sin(phi);
rO_ = [xO, yO , 0];
psi = acos((e-xO)/R); % angle between OC and x-axis
% position of C (rC_ = AC_)
xC = e;  yC = yO+R*sin(psi) ;
rC_ = [xC, yC, 0];
% Graphic of the mechanism
axis manual
axis equal
hold on
grid on
ds = .25;
axis([-ds ds -ds ds])
xlabel('x (m)'),ylabel('y (m)')
plot([0,xO],[0,yO],'k-o',[xO,xC],[yO,yC],'b-o',...
[0,xC],[0,yC],'g-o',[xC,xC],[yC,yC+Lf],'r-')
hold on
ang=0:0.01:2*pi;
xp=R*cos(ang);
yp=R*sin(ang);
ci = plot(xO+xp,yO+yp,'k--');
xm=Rc*cos(ang);
ym=Rc*sin(ang);
cm = plot(xO+xm,yO+ym,'k-');
pivot0_scale = 0.005;
plot_pivot0(0,0,pivot0_scale,0);
plot_wall(xC-.005,.2,.05,0);
```

```
plot_wall(xC+.005,.2,.05,pi);
text(0,0,'   A')
text(xO,yO,'   O')
text(xC,yC+.005,'   C')
```

The numerical results for the position analysis are

```
% phi  =  60.000 (deg)
% rO_  =  [ 0.025,  0.043,0] (m)
% psi  =  81.585 (deg)
% rC_  =  [ 0.040,  0.145,0] (m)
```

The MATLAB representation of the mechanism is shown in Fig. 8.2.

The angular velocity of the cam is $\omega_1 = \omega_1 \mathbf{k} = \pi n/30 \, \mathbf{k}$ rad/s. The angular acceleration of the cam is $\alpha_1 = d\,\omega_1/dt = \mathbf{0}$. The velocity of the point C_1 on the cam is

$$\mathbf{v}_{C_1} = \mathbf{v}_A + \omega_1 \times \mathbf{r}_C = \mathbf{0} + \begin{vmatrix} \mathbf{1} & \mathbf{j} & \mathbf{k} \\ 0 & 0 & \omega_1 \\ x_C & y_C & 0 \end{vmatrix}.$$

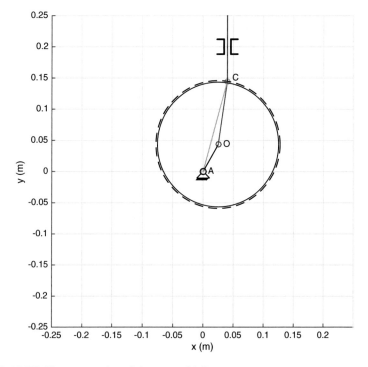

Fig. 8.2 MATLAB representation of the cam and follower

The acceleration of the point C_1 on the cam is

$$\mathbf{a}_{C_1} = \mathbf{a}_A + \boldsymbol{\alpha}_1 \times \mathbf{r}_C - \omega_1^2 \mathbf{r}_C = \mathbf{0} + \begin{vmatrix} \mathbf{1} & \mathbf{J} & \mathbf{k} \\ 0 & 0 & \alpha_1 \\ x_C & y_C & 0 \end{vmatrix} - \omega_1^2 (x_C \mathbf{1} + y_C \mathbf{J}).$$

The MATLAB commands for the velocity and acceleration of C_1 are:

```
n = 30; % (rpm)   driver link
omega1_ = [0 0 pi*n/30];
alpha1_ = [0 0 0];
vA_ = [0 0 0];
aA_ = [0 0 0];
vC1_ = vA_ + cross(omega1_,rC_);
aC1_ = aA_ + cross(alpha1_,rC_) - dot(omega1_,omega1_)*rC_;
```

and the results are:

```
% vC1_ = [-0.455, 0.126,0] (m/s)
% aC1_ = [-0.395,-1.428,0] (m/s^2)
```

The velocity of the point C_2 on the follower 2 is parallel to the vertical sliding direction

$$\mathbf{v}_{C_2} = v_{C_2} \mathbf{J}. \tag{8.1}$$

The velocity of C_2 on the follower 2 can be written in terms of the velocity of C_1 on cam 1

$$\mathbf{v}_{C_2} = \mathbf{v}_{C_1} + \mathbf{v}_{C_2 C_1} = \mathbf{v}_{C_1} + \mathbf{v}_{C_{21}}, \tag{8.2}$$

where the relative velocity $\mathbf{v}_{C_{21}}$ is along the tangential direction, Fig. 8.1b,

$$\mathbf{v}_{C_{21}} = v_{C_{21}} \mathbf{t}, \tag{8.3}$$

where the tangential unit vector is

$$\mathbf{t} = -\sin\psi \, \mathbf{1} + \cos\psi \, \mathbf{J}. \tag{8.4}$$

The normal unit vector as shown in Fig. 8.1b has the following expression

$$\mathbf{n} = -\cos\psi \, \mathbf{1} - \sin\psi \, \mathbf{J}. \tag{8.5}$$

A vectorial equation can be obtained with Eqs. (8.1)–(8.4)

$$v_{C_2} \mathbf{J} = \mathbf{v}_{C_1} + v_{C_{21}} \mathbf{t} = \mathbf{v}_{C_1} + v_{C_{21}} (-\sin \psi \mathbf{1} + \cos \psi \mathbf{J}). \tag{8.6}$$

Equation (8.6) projected on the Cartesian axis results in two scalar equations with two unknowns v_{C_2} and $v_{C_{21}}$. The MATLAB relations for the velocity of C_2 are

```
n_  = [-cos(psi), -sin(psi), 0];
t_  = [-sin(psi),  cos(psi), 0];
vC2  = sym('vC2','real');   % vC2   unknown
vC21 = sym('vC21','real');  % vC21 unknown
vC2_  = [0, vC2, 0]; % parallel to the vertical sliding direction
vC21_ = vC21*t_;   % parallel to the common tangent
eqvC = vC2_ - (vC1_ + vC21_);
eqvCx = eqvC(1); % equation component on x-axis
eqvCy = eqvC(2); % equation component on y-axis
solvC = solve(eqvCx,eqvCy);
vC2s = eval(solvC.vC2);
vC21s = eval(solvC.vC21);
vC2_  = [0, vC2s, 0];
vC21_ = vC21s*t_;
```

The numerical results are

```
% vC2_  = [0, 0.058, 0] (m/s)
% vC21_ = [0.455,-0.067, 0] (m/s)
```

The acceleration of the point C_2 on the follower 2 is parallel to the vertical sliding direction

$$\mathbf{a}_{C_2} = a_{C_2} \mathbf{J}. \tag{8.7}$$

The acceleration of C_2 on the follower 2 can be written in terms of the acceleration of C_1 on cam 1

$$\mathbf{a}_{C_2} = \mathbf{a}_{C_1} + \mathbf{a}_{C_2 C_1} = \mathbf{a}_{C_1} + \mathbf{a}_{C_{21}}^{\text{cor}} + \mathbf{a}_{C_{21}}, \tag{8.8}$$

where the Coriolis acceleration is

$$\mathbf{a}_{C_{21}}^{\text{cor}} = 2\,\boldsymbol{\omega}_1 \times \mathbf{v}_{C_{21}}, \tag{8.9}$$

and the relative acceleration $\mathbf{a}_{C_{21}}$ is along the normal and tangential directions

$$\mathbf{a}_{C_{21}} = a_{C_{21}}^n \, \mathbf{n} + a_{C_{21}}^t \, \mathbf{t}. \tag{8.10}$$

The relative normal acceleration is calculated with

$$a_{C_{21}}^n = v_{C_{21}}^2 / R. \tag{8.11}$$

From Eqs. (8.7)–(8.11) the unknowns a_{C_2} and $a_{C_{21}}^t$ are calculated. The MATLAB commands for the acceleration are:

```
aC21cor_ = 2*cross(omega1_, vC21_);
aC21n_ = (dot(vC21_, vC21_)/R)*n_;
aC2  = sym('aC2','real');   % aC2   unknown
aC21t = sym('aC21t','real');  % aC21t unknown
aC2_ = [0, aC2, 0]; % aC2 parallel to sliding direction y axis
aC21t_ = aC21t*t_; % aC21t_ parallel to the common tangent
eqaC = aC2_ - (aC1_ + aC21n_ + aC21t_ + aC21cor_);
eqaCx = eqaC(1); % equation component on x-axis
eqaCy = eqaC(2); % equation component on y-axis
solaC = solve(eqaCx,eqaCy);
aC2s = eval(solaC.aC2);
aC21ts = eval(solaC.aC21t);
aC2_ = [0, aC2s, 0];
aC21t_ = aC21ts*t_;
```

and the numerical results are:

```
% aC21cor_ = [0.423, 2.856, 0] (m/s^2)
% aC21n_ = [-0.301, -2.038, 0] (m/s^2)
% aC2_ = [0, -0.650, 0] (m/s^2)
% aC21t_ = [0.274, -0.040, 0] (m/s^2)
```

8.2 Force Analysis

The external force of magnitude $F_e = 100$ N is acting on the follower as shown in Figs. 8.1a and 8.3

$$\mathbf{F}_e = -F_e \mathbf{J}. \tag{8.12}$$

The mass of the follower is given by

$$m_2 = \rho L_f w d + \rho (\pi R_f^2 / 2) d,$$

and the weight of the follower is

$$\mathbf{G}_2 = -m_2 g \mathbf{J}.$$

The reaction force of the link 1 on link 2 is

Fig. 8.3 Force analysis for cam and follower

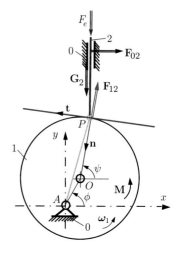

$$\mathbf{F}_{12} = F_{12}\cos\psi\,\mathbf{i} + F_{12}\sin\psi\,\mathbf{j}.$$

The reaction force of the ground 0 on link 2 is

$$\mathbf{F}_{02} = F_{02}\,\mathbf{i}.$$

The Newton's equation of motion for the follower is

$$m_2\,\mathbf{a}_{C_2} = \mathbf{F}_{02} + \mathbf{F}_{12} + \mathbf{F}_e + \mathbf{G}_2. \qquad (8.13)$$

From Eq. (8.13) the unknowns F_{02} and F_{12} are calculated. The MATLAB relations for the follower are:

```
Fe_ = [0, -Fe, 0];
m2 = rho*Lf*w*d + rho*(pi*Rf^2/2)*d;
G2_ = [0, -m2*g, 0];
F02=sym('F02','real');
F12=sym('F12','real');
F02_ = [F02, 0, 0];
F12_ = [F12*cos(psi), F12*sin(psi),0];
% m2*aC2_ = F02_ + F12_ + G2_ + Fe_
eqF2 = F02_ + F12_ + G2_ + Fe_ - m2*aC2_;
solF2=solve(eqF2(1),eqF2(2));
F02s=eval(solF2.F02);
F12s=eval(solF2.F12);
F12_ = [F12s*cos(psi), F12s*sin(psi),0];
```

and the results are:

```
% m2  = 0.0608 (kg)
% F02 = -14.876 (N)
% F12 = 101.651 (N)
% F12_ = [14.876, 100.557, 0] (N)
```

The equation of motion for the rotating cam 1 is given by

$$I_A \, \alpha_1 = \mathbf{M}_m + \mathbf{r}_O \times \mathbf{G}_1 + \mathbf{r}_P \times \mathbf{F}_{21}, \qquad (8.14)$$

where I_A is the mass moment of inertia of the cam about the fixed point A, $\alpha_1 = 0$, $\mathbf{F}_{21} = -\mathbf{F}_{12}$, $\mathbf{G}_1 = -m_1 \, g \, \mathbf{J}$, $m_1 = \rho \, \pi \, R_c^2 \, d$ is the mass of the cam, and $\mathbf{r}_P = \mathbf{r}_{AP}$ is the position vector of the contact point P shown in Fig. 8.1a

$$\mathbf{r}_P = (e - R_f \, \cos \psi) \mathbf{\imath} + (y_O + R_c \, \sin \psi \, \mathbf{J}).$$

The dynamic equilibrium moment $\mathbf{M}_m$ is calculated as

$$\mathbf{M}_m = -\mathbf{r}_O \times \mathbf{G}_1 - \mathbf{r}_P \times \mathbf{F}_{21}.$$

The MATLAB relations for the cam are:

```
m1 = rho*pi*Rc^2*d;
G1_ = [0, -m1*g, 0];
% IA*alpha1_ = Mm_+rO_xG1_+rP_xF21_
% P contact point between the cam and follower
xP = e - Rf*cos(psi);
yP = yO + Rc*sin(psi);
rP_ = [xP, yP, 0];
Mm_ = -cross(rO_, G1_)-cross(rP_, -F12_);
```

and the results are

```
% m1 = 2.5133 (kg)
% Mm = 2.4860 (N m)
```

8.3 Equivalent Linkages

Between the follower and the cam there is a two degrees of freedom joint. A two degrees of freedom joint is usually referred to as a half-joint and has 4 degrees of constraint. A two degrees of freedom joint can be kinematically replaced with two one degree of freedom joints and a link. The cam joint can be replaced with two full-joints (one degree of freedom joints) at O and C and a link 3 as shown in Fig. 8.4. The cam and follower mechanism is kinematically equivalent with the mechanism

Fig. 8.4 Equivalent linkages
for the cam and follower

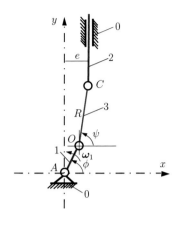

in Fig. 8.4. The velocity and acceleration of the point O on the link 1, Fig. 8.4,
are

$$\mathbf{v}_O = \mathbf{v}_A + \boldsymbol{\omega}_1 \times \mathbf{r}_O,$$
$$\mathbf{a}_O = \mathbf{a}_A + \boldsymbol{\alpha}_1 \times \mathbf{r}_O - \omega_1^2 \mathbf{r}_O, \tag{8.15}$$

where $\omega_1 = \pi n/30 \, \mathbf{k}$ rad/s and $\alpha_1 = d \, \omega_1/dt = \mathbf{0}$. The velocity and acceleration
of the point $C = C_3$ on link 3 are

$$\mathbf{v}_C = \mathbf{v}_{C_3} = \mathbf{v}_O + \boldsymbol{\omega}_3 \times \mathbf{r}_{OC},$$
$$\mathbf{a}_C = \mathbf{a}_{C_3} = \mathbf{a}_O + \boldsymbol{\alpha}_3 \times \mathbf{r}_{OC} - \omega_3^2 \mathbf{r}_{OC}, \tag{8.16}$$

where $\boldsymbol{\omega}_3 = \omega_3 \, \mathbf{k}$, $\boldsymbol{\alpha}_3 = \alpha_3 \, \mathbf{k}$, and

$$\mathbf{v}_C = \mathbf{v}_{C_2} = v_C \, \mathbf{J} \quad \text{and} \quad \mathbf{a}_C = \mathbf{a}_{C_2} = a_C \, \mathbf{J}. \tag{8.17}$$

From Eqs. (8.15)–(8.17) the unknowns v_C, a_C, ω_3, and α_3 are calculated. The MAT-
LAB program for the equivalent linkages is:

```
n = 30; % (rpm)   driver link
omega1_ = [ 0 0 pi*n/30 ];
alpha1_ = [0 0 0 ];
vA_ = [0 0 0 ];   aA_ = [0 0 0 ];
vO_ = vA_ + cross(omega1_,rO_);
aO_ = aA_ + cross(alpha1_,rO_) - dot(omega1_,omega1_)*rO_;
% or
aO_ = aA_ + cross(alpha1_,rO_) +...
    cross(omega1_,cross(omega1_,rO_));
vC  = sym('vC','real'); % vC2  unknown
```

```
omega3 = sym('omega3','real'); % omega3 unknown
% vC2 || yy
vC2_ = [0,vC,0];
omega3_ = [0,0,omega3];
% vC3_ = vO_ + omega3_ x OC_
vC3_ = vO_+cross(omega3_,rC_-rO_);
% vC_=vC2_=vC3_
eqvC = vC2_ - vC3_;
eqvCx = eqvC(1); % equation component on x-axis
eqvCy = eqvC(2); % equation component on y-axis
solvC = solve(eqvCx,eqvCy);
vCs = eval(solvC.vC);
omega3s = eval(solvC.omega3);
vC_ = [0,vCs,0];
omega3_ = [0,0,omega3s];
aC = sym('aC','real'); % aC2 unknown
alpha3 = sym('alpha3','real'); % alpha3 unknown
% vC2 || yy
aC2_ = [0,aC,0];
alpha3_ = [0,0,alpha3];
% aC3_ = aO_ + alpha3_ x OC_ - omega3^2 OC_
aC3_ = aO_+cross(alpha3_,rC_-rO_)-dot(omega3_,omega3_)*(rC_-rO_);
eqaC = aC2_ - aC3_;
eqaCx = eqaC(1); % equation component on x-axis
eqaCy = eqaC(2); % equation component on y-axis
solaC = solve(eqaCx,eqaCy);
aCs = eval(solaC.aC);
alpha3s = eval(solaC.alpha3);
aC_ = [0,aCs,0];
alpha3_ = [0,0,alpha3s];
```

and the results are:

```
% vC_ = [0, 0.058,0] (m/s)
% omega3_ = [0,0,-1.342] (rad/s)
% aC_ = [0,-0.650,0] (m/s^2)
% alpha3_ = [0,0,-2.700] (rad/s^2)
```

8.4   Differential Method

The position of the center O, Fig. 8.1, is given by

$$x_O(t) = x_A + AO \cos \phi(t) = AO \cos \phi(t),$$
$$y_O(t) = y_A + AO \sin \phi(t) = AO \sin \phi(t),$$

where $\phi = \phi(t)$ is a function of time. The position of the point C is calculated with

$$x_C(t) = e,$$
$$y_C(t) = y_O + R \sin \psi,$$

where $\psi = \arccos\,[(e - x_O)/R]$ and $R = R_c + R_f$. The linear velocity vector of $C = C_2$ on the cam is

$$\mathbf{v}_C = \mathbf{v}_{C_2} = \dot{x}_C \mathbf{1} + \dot{y}_C \mathbf{J},$$

and the linear acceleration vector of $C = C_2$ is

$$\mathbf{a}_C = \mathbf{a}_{C_2} = \ddot{x}_C \mathbf{1} + \ddot{y}_C \mathbf{J}.$$

The MATLAB commands for calculating the velocity and the acceleration of the follower are:

```
syms phi(t)

% position of O = Oc (rO_ = AO_)
xO = RO*cos(phi);
yO = RO*sin(phi);
rO_ = [xO, yO , 0];


% angle between OC and x-axis
psi = acos((e-xO)/R);

% position of C (rC_ = AC_)
xC = e;
yC = yO+R*sin(psi) ;
rC_ = [xC, yC, 0];


vC2_ = diff(rC_, t);
aC2_ = diff(vC2_,t);


n = 30; % (rpm)  driver link
omega1 = pi*n/30;
alpha1 = 0;
phi1   = pi/3;
```

```
slist={diff(phi(t),t,2), diff(phi(t),t), phi(t)};
nlist={alpha1,  omega1, phi1}; %numerical values for slist

vC2n_ = eval(subs(vC2_,slist,nlist));
aC2n_ = eval(subs(aC2_,slist,nlist));
```

and the numerical results are:

```
% vC2_ = [0,   0.058, 0] (m/s)
% aC2_ = [0, -0.650, 0] (m/s^2)
```

8.5 Problems

8.1 The prism ABC shown in Fig. 8.5 is the link 1 and has the angle $\angle BAC = \phi$. The prism 1 has a translational motion and at the time t has the speed u and the acceleration a. The bar ED is the link 2 and is resting on the prism at D. The link 2 is moving on a perpendicular direction to AB. Find the velocity and the acceleration of the point D on link 2 at the time t. For the numerical application use: $\phi = 45°$, $u = 1$ m/s, and $a = 2$ m/s^2.

8.2 The prism 1 in Fig. 8.6 has a translational motion and at the time t the velocity of link 1 is $v_1 = u$ and the acceleration of 1 is $a_1 = u^2/l$. The following data are given: $\angle ACB = \phi$ and $OD = l$. At the time t the angle of the link 2 is $\angle(OD, OE) = \phi$, where $OE||BC$. Find the angular velocity and acceleration of the link 2 at the time t. For the numerical application use: $\phi = \pi/6, l = 0.1$ m, and $u = 1$ m/s.

Fig. 8.5 Problem 8.1

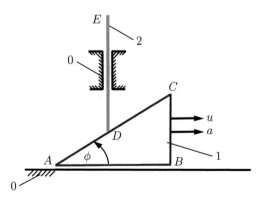

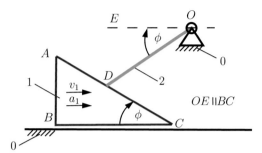

Fig. 8.6 Problem 8.2

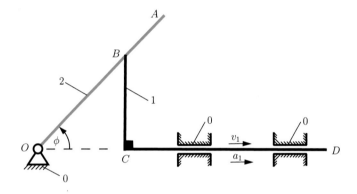

Fig. 8.7 Problem 8.3

8.3 For the mechanism in Fig. 8.7, the following data are given: $BC = l$, $OA = 2l$, $\angle BCD = 90°$. For the position shown in the figure $\angle BOC = \phi$ and the link 1 has a translational motion with the velocity $v_1 = u$ and with the acceleration $a_1 = 2u^2/l$. Find the angular velocity and acceleration of the link 2 for the given position. For the numerical application use: $l = 20$ cm, $\phi = 45°$, and $u = 40$ cm/s.

8.4 The curved link 3, in Fig. 8.8, has the radius $R = 0.3$ m, a counterclockwise angular velocity of 2 rad/s, and a clockwise angular acceleration of 6 rad/s^2 when the angle $\phi = 30°$. For this configuration determine the velocities and the accelerations of the mechanism using two methods.

8.5 The disk 3, shown in Fig. 8.9, with the radius $R = 0.4$ m, has a constant angular speed $n = 20$ rpm, clockwise direction. Calculate the velocities and the accelerations of the mechanism, using two methods, when $\phi = 45°$.

Fig. 8.8 Problem 8.4

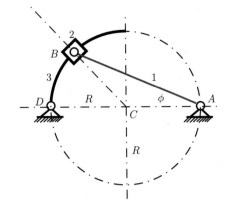

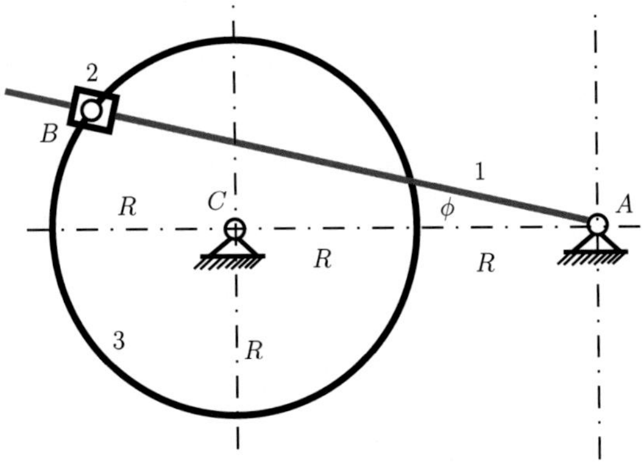

Fig. 8.9 Problem 8.5

References

1. E.A. Avallone, T. Baumeister, A. Sadegh, *Marks' Standard Handbook for Mechanical Engineers*, 11th edn. (McGraw-Hill Education, New York, 2007)
2. I.I. Artobolevski, *Mechanisms in Modern Engineering Design* (MIR, Moscow, 1977)
3. M. Atanasiu, *Mechanics (Mecanica)* (EDP, Bucharest, 1973)
4. H. Baruh, *Analytical Dynamics* (WCB/McGraw-Hill, Boston, 1999)
5. A. Bedford, W. Fowler, *Dynamics* (Addison Wesley, Menlo Park, CA, 1999)
6. A. Bedford, W. Fowler, *Statics* (Addison Wesley, Menlo Park, CA, 1999)
7. F.P. Beer et al., *Vector Mechanics for Engineers: Statics and Dynamics* (McGraw-Hill, New York, NY, 2016)
8. M. Buculei, D. Bagnaru, G. Nanu, D.B. Marghitu, *Computing Methods in the Analysis of the Mechanisms with Bars* (Scrisul Romanesc, Craiova, 1986)
9. M.I. Buculei, *Mechanisms* (University of Craiova Press, Craiova, Romania, 1976)
10. D. Bolcu, S. Rizescu, *Mecanica* (EDP, Bucharest, 2001)
11. J. Billingsley, *Essential of Dynamics and Vibration* (Springer, 2018)
12. R. Budynas, K.J. Nisbett, *Shigley's Mechanical Engineering Design*, 9th edn. (McGraw-Hill, New York, 2013)
13. J.A. Collins, H.R. Busby, G.H. Staab, *Mechanical Design of Machine Elements and Machines*, 2nd edn. (Wiley, 2009)

14. M. Crespo da Silva, *Intermediate Dynamics for Engineers* (McGraw-Hill, New York, 2004)
15. A.G. Erdman, G.N. Sandor, *Mechanisms Design* (Prentice-Hall, Upper Saddle River, NJ, 1984)
16. M. Dupac, D.B. Marghitu, *Engineering Applications: Analytical and Numerical Calculation with MATLAB* (Wiley, Hoboken, NJ, 2021)
17. A. Ertas, J.C. Jones, *The Engineering Design Process* (Wiley, New York, 1996)
18. F. Freudenstein, An application of Boolean Algebra to the Motion of Epicyclic Drivers. J. Eng. Ind. 176–182 (1971)
19. J.H. Ginsberg, *Advanced Engineering Dynamics* (Cambridge University Press, Cambridge, 1995)
20. H. Goldstein, *Classical Mechanics* (Addison-Wesley, Redwood City, CA, 1989)
21. D.T. Greenwood, *Principles of Dynamics* (Prentice-Hall, Englewood Cliffs, NJ, 1998)
22. A.S. Hall, A.R. Holowenko, H.G. Laughlin, *Schaum's Outline of Machine Design* (McGraw-Hill, New York, 2013)
23. B.G. Hamrock, B. Jacobson, S.R. Schmid, *Fundamentals of Machine Elements* (McGraw-Hill, New York, 1999)
24. R.C. Hibbeler, *Engineering Mechanics: Dynamics* (Prentice Hall, 2010)
25. T.E. Honein, O.M. O'Reilly, On the Gibbs–Appell Equations for the Dynamics of Rigid Bodies J. Appl. Mech. **88**/074501-1 (2021)
26. R.C. Juvinall, K.M. Marshek, *Fundamentals of Machine Component Design*, 5th edn. (Wiley, New York, 2010)
27. T.R. Kane, *Analytical Elements of Mechanics*, vol. 1 (Academic Press, New York, 1959)
28. T.R. Kane, *Analytical Elements of Mechanics*, vol. 2 (Academic Press, New York, 1961)
29. T.R. Kane, D.A. Levinson, The use of Kane's dynamical equations in robotics. MIT Int. J. Robot. Res. **3**, 3–21 (1983)
30. T.R. Kane, P.W. Likins, D.A. Levinson, *Spacecraft Dynamics* (McGraw-Hill, New York, 1983)
31. T.R. Kane, D.A. Levinson, *Dynamics* (McGraw-Hill, New York, 1985)
32. K. Lingaiah, *Machine Design Databook*, 2nd edn. (McGraw-Hill Education, New York, 2003)
33. N.I. Manolescu, F. Kovacs, A. Oranescu, *The Theory of Mechanisms and Machines* (EDP, Bucharest, 1972)
34. D.B. Marghitu, *Mechanical Engineer's Handbook* (Academic Press, San Diego, CA, 2001)
35. D.B. Marghitu, M.J. Crocker, *Analytical Elements of Mechanisms* (Cambridge University Press, Cambridge, 2001)
36. D.B. Marghitu, *Kinematic Chains and Machine Component Design* (Elsevier, Amsterdam, 2005)
37. D.B. Marghitu, *Mechanisms and Robots Analysis with MATLAB* (Springer, New York, NY, 2009)
38. D.B. Marghitu, M. Dupac, *Advanced Dynamics: Analytical and Numerical Calculations with MATLAB* (Springer, New York, NY, 2012)
39. D.B. Marghitu, M. Dupac, H.M. Nels, *Statics with MATLAB* (Springer, New York, NY, 2013)
40. D.B. Marghitu, D. Cojocaru, *Advances in Robot Design and Intelligent Control* (Springer International Publishing, Cham, Switzerland, 2016), pp. 317–325
41. D.J. McGill, W.W. King, *Engineering Mechanics: Statics and an Introduction to Dynamics* (PWS Publishing Company, Boston, 1995)
42. J.L. Meriam, L.G. Kraige, *Engineering Mechanics: Dynamics* (Wiley, New York, 2007)
43. C.R. Mischke, Prediction of Stochastic Endurance strength. Trans. ASME J. Vib. Acoust. Stress Reliab. Des. **109**(1), 113–122 (1987)
44. L. Meirovitch, *Methods of Analytical Dynamics* (Dover, 2003)
45. R.L. Mott, *Machine Elements in Mechanical Design* (Prentice Hall, Upper Saddle River, NJ, 1999)
46. W.A. Nash, *Strength of Materials* (Schaum's Outline Series, McGraw-Hill, New York, 1972)
47. R.L. Norton, *Machine Design* (Prentice-Hall, Upper Saddle River, NJ, 1996)
48. R.L. Norton, *Design of Machinery* (McGraw-Hill, New York, 1999)
49. O.M. O'Reilly, *Intermediate Dynamics for Engineers Newton-Euler and Lagrangian Mechanics* (Cambridge University Press, UK, 2020)

50. O.M. O'Reilly, *Engineering Dynamics: A Primer* (Springer, NY, 2010)
51. W.C. Orthwein, *Machine Component Design* (West Publishing Company, St. Paul, 1990)
52. L.A. Pars, *A Treatise on Analytical Dynamics* (Wiley, New York, 1965)
53. F. Reuleaux, *The Kinematics of Machinery* (Dover, New York, 1963)
54. D. Planchard, M. Planchard, *SolidWorks 2013 Tutorial with Video Instruction* (SDC Publications, 2013)
55. I. Popescu, *Mechanisms* (University of Craiova Press, Craiova, Romania, 1990)
56. I. Popescu, C. Ungureanu, *Structural Synthesis and Kinematics of Mechanisms with Bars* (Universitaria Press, Craiova, Romania, 2000)
57. I. Popescu, L. Luca, M. Cherciu, D.B. Marghitu, *Mechanisms for Generating Mathematical Curves* (Springer Nature, Switzerland, 2020)
58. C.A. Rubin, *The Student Edition of Working Model* (Addison-Wesley Publishing Company, Reading, MA, 1995)
59. J. Ragan, D.B. Marghitu, Impact of a Kinematic link with MATLAB and SolidWorks. Applied Mechanics and Materials **430**, 170–177 (2013)
60. J. Ragan, D.B. Marghitu, MATLAB dynamics of a free link with elastic impact, in *International Conference on Mechanical Engineering, ICOME 2013*, 16–17 May 2013 (Craiova, Romania, 2013)
61. J.C. Samin, P. Fisette, *Symbolic Modeling of Multibody Systems* (Kluwer, 2003)
62. A.A. Shabana, *Computational Dynamics* (Wiley, New York, 2010)
63. I.H. Shames, *Engineering Mechanics—Statics and Dynamics* (Prentice-Hall, Upper Saddle River, NJ, 1997)
64. J.E. Shigley, C.R. Mischke, *Mechanical Engineering Design* (McGraw-Hill, New York, 1989)
65. J.E. Shigley, C.R. Mischke, R.G. Budynas, *Mechanical Engineering Design*, 7th edn. (McGraw-Hill, New York, 2004)
66. J.E. Shigley, J.J. Uicker, *Theory of Machines and Mechanisms* (McGraw-Hill, New York, 1995)
67. D. Smith, *Engineering Computation with MATLAB* (Pearson Education, Upper Saddle River, NJ, 2008)
68. R.W. Soutas-Little, D.J. Inman, *Engineering Mechanics: Statics and Dynamics* (Prentice-Hall, Upper Saddle River, NJ, 1999)
69. J. Sticklen, M.T. Eskil, *An Introduction to Technical Problem Solving with MATLAB* (Great Lakes Press, Wildwood, MO, 2006)
70. A.C. Ugural, *Mechanical Design* (McGraw-Hill, New York, 2004)
71. R. Voinea, D. Voiculescu, V. Ceausu, *Mechanics (Mecanica)* (EDP, Bucharest, 1983)
72. K.J. Waldron, G.L. Kinzel, *Kinematics, Dynamics, and Design of Machinery* (Wiley, New York, 1999)
73. J.H. Williams Jr., *Fundamentals of Applied Dynamics* (Wiley, New York, 1996)
74. C.E. Wilson, J.P. Sadler, *Kinematics and Dynamics of Machinery* (Harper Collins College Publishers, New York, 1991)
75. H.B. Wilson, L.H. Turcotte, D. Halpern, *Advanced Mathematics and Mechanics Applications Using MATLAB* (Chapman & Hall/CRC, 2003)
76. S. Wolfram, *Mathematica* (Wolfram Media/Cambridge University Press, Cambridge, 1999)
77. National Council of Examiners for Engineering and Surveying (NCEES), *Fundamentals of Engineering. Supplied-Reference Handbook* (Clemson, SC, 2001)
78. * * *, *The Theory of Mechanisms and Machines* (Teoria mehanizmov i masin) (Vassaia scola, Minsk, Russia, 1970)
79. eCourses—University of Oklahoma. http://ecourses.ou.edu/home.htm
80. https://www.mathworks.com
81. http://www.eng.auburn.edu/~marghitu/
82. https://www.solidworks.com
83. https://www.wolfram.com

Chapter 9
Direct Dynamics

Abstract The direct dynamics of a body in planar motion is studied. The equations of motion are developed with symbolic MATLAB. At a certain moment an external elastic force acts on the body. MATLAB ODE solver is employed to integrate the equations taking into consideration different events. The impact of a free kinematic link with a surface is also studied. The results are obtained with Newton-Euler and Lagrange methods.

9.1 Equations of Motion—Sphere on a Spring

In classical mechanics it was admitted the continuity of velocities, of accelerations and forces. The mechanical phenomena are continuous in time. There are cases when for very small time intervals there is a very large variation of velocity, but the displacement in space of the particle is very small.

If a sphere falls on a vertical spring, the sphere makes contact with the spring. The spring compresses under the weight of the sphere. The compression phase ends when the velocity of the sphere is zero. Next phase is the restitution phase when the spring is expanding and the sphere is moving upward. At the end of the restitution phase there is the separation of the sphere.

The x-axis is downward as shown in the Fig. 9.1. Assume that at moment $t = 0$ the sphere gets in contact with the spring. At that moment the velocity of the sphere is:

$$\mathbf{v}(t = 0) = \mathbf{v}_0 = v_0\,\boldsymbol{\imath}.$$

The equation of motion for the sphere in contact with the spring is:

$$m\,\ddot{x} = m\,g - k\,x, \tag{9.1}$$

where k is the spring constant ($k > 0$).

Fig. 9.1 Sphere falls on a
vertical spring

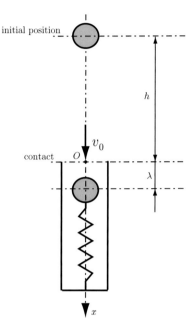

The contact force is due to elastic force:

$$\mathbf{P} = -k\,x\,\mathbf{1}.$$

The initial conditions are

$$x(0) = 0, \ \dot{x}(0) = v_0.$$

The symbolic equation of motion is solved with dsolve:

```
syms x(t)
syms m k g h xo vo
assume(m,'positive') % mass of the small sphere
assume(k,'positive') % spring constant
assume(g,'positive') % gravitational acceleration
assume(xo,'positive') % intial condition for displacement
assume(vo,'positive') % intial condition for velocity

% equation of motion
% m*diff(x,2) == m*g-k*x
eqn = m*diff(x,2) + k*x - m*g;
v = diff(x);
% dsolve solves symbolic differential equation
```

```
xSol = dsolve(eqn, x(0)==xo, v(0)==vo);
vSol = diff(xSol,t);
fprintf('x(t)=\n')
pretty(xSol)
fprintf('v(t)=\n')
pretty(vSol)
fprintf('\n')
```

The symbolic solution $x(t)$ and $\dot{x}(t) = v(t)$ are:

```
% x(t)=
%                / sqrt(k) t \                            / sqrt(k) t \
%           cos| --------- | (g m - k xo)   sqrt(m) vo sin| --------- |
% g m           \ sqrt(m) /                               \ sqrt(m) /
% --- - ----------------------------- + -----------------------------
% k                     k                            sqrt(k)
%
% v(t)=
%                           / sqrt(k) t \
%                      sin| --------- | (g m - k xo)
%           / sqrt(k) t \   \ sqrt(m) /
% vo cos| --------- | + -----------------------------
%           \ sqrt(m) /            sqrt(k) sqrt(m)
```

The numerical input data for the system are:

```
% numerical values
lists = { m, g, k, h}; % symbolic list
listn = {10, 9.81, 294e3, 1}; % numerical list
% m = 10 (kg) mass
% g = 9.81 (m/s^2) gravitational acceleration
% k = 294e3 (N/m) spring constant
% h = 1 (m) drop height
% v0 = sqrt(2*g*h) initial velocity (m/s)
```

The initial conditions at $t(0) = t_0 = 0$ are:

```
% initial conditions at t0=0 s
x0 = 0; % (m) x(0) = 0
% contact velocity v(0) = (2*g*h)^(1/2)
v0 = subs(sqrt(2*g*h), lists, listn); % (m/s)
% v0 = 4.429 (m/s)
```

Substituting the numerical data for the functions in time for positions and velocities are:

```
xn = subs(xSol, lists, listn);
xn = subs(xn, {xo,vo}, {x0, v0});
```

```
xn = vpa(xn,3);

vn = subs(vSol, lists, listn);
vn = subs(vn, {xo,vo}, {x0, v0});
vn = vpa(vn,3);

fprintf('x(t)=\n')
pretty(xn)
fprintf('v(t)=\n')
pretty(vn)
fprintf('\n')

% x(t)=
% sin(171.0 t) 0.0258 - cos(171.0 t) 3.34e-4 + 3.34e-4
%
% v(t)=
% cos(171.0 t) 4.43 + sin(171.0 t) 0.0572
```

The sphere reaches its maximum position on $x-$axis at $t = t_v$ when $\dot{x}(t_v) = v(t_v) = 0$, or in MATLAB:

```
t0 = 0;      % intial time
t2 = 0.02;   % arbitrary select a final time
% solve the equation v(t)=0 => t = tv
% v(t)=dx(t)/dt=0 => x(t)= x maximum
tv = vpasolve(vn, t, [t0 t2]);
tv = double(tv);
fprintf('tv = %6.4f (s)\n',tv)
% tv = 0.0092 (s)
```

The final time t_f is obtained from the equation $x(t) = 0$. At the moment $t = t_f$, the sphere attains again the reference O and the sphere separates itself and moves upward:

```
% solve the equation x(t)=0 => t = tf > tv
tf = vpasolve(xn, t, [tv t2]);
tf = double(tf);
fprintf('tf = %6.4f (s)\n',tf)
% tf = 0.0185 (s)  % end of contact
```

The jump in velocity between the two moments of contact and separation is

```
% jump in velocity
deltav = v0-vf;
% v0-vf = 8.859 (m/s)
```

The contact elastic force function is:

```
k = 294e3; % spring constant (N/m)
Pen = k*xn; % (N) contact elastic force
```

The maximum elastic displacement and the maximum elastic contact force are

```
xmax = subs(xn, t, tv); % maximum displacement at tv
Pmax = subs(Pen,t, tv); % maximum contact elastic force at tv
% x max =   0.026 (m)
% P max = 7693.653 (N)
```

The maximum elastic force is actually much greater than the weight of the sphere. The displacement of the sphere is very small while the velocity's jump is big. The MATLAB results are depicted in Fig. 9.2. Figure 9.2a represents the dependence of the displacement of the sphere with respect to time and Fig. 9.2b shows the variation in time of the velocity of the sphere in contact with the spring. At $t = 0$ the sphere gets in contact with the spring, and at $t = t_f$, the sphere separates from the spring. Note that the initial velocity is equal with the absolute value of the final velocity. Figure 9.2c represents the variation of the elastic force in time, with its maximum at the time of maximum compression of the spring. The code for the plots is:

```
figure
subplot(3,1,1)
fplot(xn, [0 tf],'-k','Linewidth',2)
ylabel('x (m)')
xlabel('time t (s)')
grid on
subplot(3,1,2)
fplot(vn, [0 tf],'-k','Linewidth',2)
ylabel('dx/dt (m/s)')
xlabel('time t (s)')
grid on
subplot(3,1,3)
fplot(Pen, [0 tf],'-k','Linewidth',2)
ylabel('Pe (N)'), xlabel('t (s)')
grid on
```

Next the differential equation of motion given by Eq. (9.1) is solved numerically using the MATLAB function ode45. Numerical approaches have the advantage of being simple to apply for complex mechanical systems. The differential equation $m\ddot{x} = m g - k x$ is of order 2. The second order equation has to be re-written as a first-order system. Let $x_1 = x$ and $x_2 = \dot{x}$, this gives the first-order system

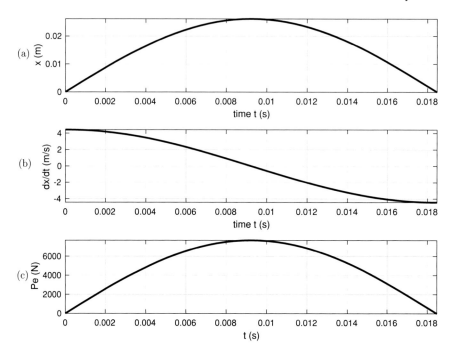

Fig. 9.2 MATLAB simulations: **a** displacement of the sphere, **b** velocity of the sphere, and **c** elastic force

$$\dot{x}_1 = x_2,$$
$$\dot{x}_2 = g - \frac{k}{m} x_1.$$

The MATLAB commands for the right-hand side of the first-order system are:

```
syms t x1 x2
dx1 = x2;
dx2 = g-k*x1/m;
```

where `syms` creates symbolic numbers, variables and objects. Using the command `matlabFunction(dx,'file','mydx','vars',t, [x1; x2])` a MATLAB file or anonymous function is obtained from a symbolic object:

```
dx = [dx1; dx2];
g = matlabFunction(dx,'vars',{t, [x1; x2]});
```

A function `eventx1.m` is constructed to stop the integration when $x = x_1 = 0$:

```
function [value,isterminal,direction] = eventx1(t,x)
% y  = x(1);
% y' = x(2);
value = x(1);
isterminal = 1;
direction = 0;
% if isterminal vector is set to 1, integration will stop
% when a zero-crossing is detected.
% The elements of the direction vector are -1, 1, or 0,
% specifying that the corresponding event must be decreasing,
% increasing, or that any crossing is to be detected
```

The function eventx1.m is used for the ODE option to find the case when $x = x_1 = 0$:

```
%options for ODE solver
optionx1 = odeset('RelTol',1e-6,'MaxStep',1e-6,...
        'Events',@eventx1);
% odeset creates ODE options structure
% RelTol - relative error tolerance
% MaxStep - upper bound on step size
% Events - detects events
```

The initial conditions at $t_0 = 0$ are $x(0) = 0$ m and $\dot{x}(0) = v_0$ m/s or in MATLAB: ic0 = [0 v0]. The time t is going from an initial value t0 to a final arbitrary value tf: t0 = 0; tf = 1; ti = [t0 tf]. The numerical solution is obtained using the command:

```
[tte, xse, te, ye] = ode45(g, ti, ic0, optionx1);
```

The integration is stopped at the moment tf = te(2) when the velocity of the sphere is vf = ye(4).

With the ode45 command the time vector and the solutions are obtained.

```
time = tte;
dis = xse(:,1);
vel = xse(:,2);
```

The x vector displacement values are xse(:,1) and the velocities are xse(:,2). The maximum displacement and the maximum force are:

```
Pen = k*dis;
xmax = max(dis);
Pmax = max(Pen);
```

The plot of the solution curves are obtained using the commands:

```
figure(1)
grid on
subplot(3,1,1)
plot(time,dis,'LineWidth',1.5)
ylabel('x (m)')
grid on
subplot(3,1,2)
xlabel('t (s)')
plot(time,vel,'LineWidth',1.5)
ylabel('dx/dt (m/s)')
grid on
subplot(3,1,3)
plot(time,Pen,'LineWidth',1.5)
ylabel('Pe (N)'), xlabel('t (s)')
grid on
```

The results obtained with the numerical approach are identical with the results obtained with the analytical method.
The complete MATLAB program is:

```
clear; clc; close all
g = 9.81; % gravitational acceleration (m/s^2)
m = 10; % mass (kg)
k1 = 294e3; % spring constant (N/m)
h = 1; % drop height (m)
v0 = sqrt(2*g*h); % initial velocity (m/s)
% equation of motion
% m x'' = m g-k*x
syms t x1 x2
dx1 = x2;
dx2 = g-k*x1/m;
dx = [dx1; dx2];
g = matlabFunction(dx,'vars',{t, [x1; x2]});
% options for ODE solver
optionx1 = odeset('RelTol',1e-6,'MaxStep',1e-6,...
        'Events',@eventx1);
ic0 = [0 v0]; % initial conditions
t0 = 0; tf = 1; ti = [t0 tf]; % time
[tte, xse, te, ye] = ode45(g, ti, ic0, optionx1);
tf = te(2);
vf = ye(4);
time = tte;
dis = xse(:,1);
```

```
vel = xse(:,2);
Pen = k1*dis;
xmax = max(dis);
Pmax = max(Pen);

function [value,isterminal,direction] = eventx1(~,x)
value = x(1);
isterminal = 1;
direction = 0;
end
```

9.2   Dynamics of a Rotating Link with an Elastic Force

Figure 9.3 depicts a uniform cylindrical compound pendulum of mass m, length L, and radius R. The rotating link is connected to the ground by a pin joint and is free to rotate in a vertical plane. The instantaneous angle of the link with the horizontal axis is $\theta(t)$. The link is dropped at a horizontal position when $\theta(0) = 0$. A vertical elastic force $\mathbf{F}_e$ acts on the end point A of the pendulum when the link makes a given angle θ_0 with the horizontal axis. The elastic force is acting upward and has the magnitude $F_e = k\,|\delta|^\alpha$, where k, α are constants and δ is the elastic deformation. The motion of the pendulum is analyzed with MATLAB.

Numerical example
The pendulum has the length $L = 0.50$ m and has a diameter $d = 0.01$ m. The density of the link is $\rho = 7\,800$ kg/m^3. The elastic force $\mathbf{F}_e$ is applied when $\theta(t_0) = \theta_0 = -\pi/6$. The elastic constant is $k = 2\,000$ and $\alpha = 1.5$.

The plane of motion will be designated the xy plane. The $y-$axis is vertical, with the positive sense directed vertically upward. The $x-$axis is horizontal and is contained in the plane of motion. The $z-$axis is also horizontal and is perpendicular to the plane of motion. These axes define an inertial reference frame. The unit vectors for the inertial reference frame are $\mathbf{i}$, $\mathbf{j}$, and $\mathbf{k}$.

The mass center of the link is at the point C. As the link is uniform, its mass center is coincident with its geometric center. The mass center, C, is at a distance $L/2$ from the pivot point O fixed to the ground, and the position vector is

$$\mathbf{r}_{OC} = \mathbf{r}_C = x_C\,\mathbf{i} + y_C\,\mathbf{j},$$

where x_C and y_C are the coordinates of C

$$x_C = \frac{L}{2}\cos\theta, \quad y_C = \frac{L}{2}\sin\theta.$$

Hence

Fig. 9.3 Rotating link in
contact with an elastic force

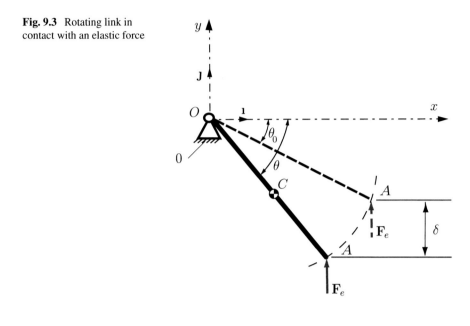

$$\mathbf{r}_C = \frac{L}{2} \cos \theta \, \mathbf{1} + \frac{L}{2} \sin \theta \, \mathbf{J}. \tag{9.2}$$

The point A at the end of the link, that gets in contact with the elastic force, is located
at the distance L from the pivot point O. The position vector of A is

$$\mathbf{r}_{OA} = \mathbf{r}_A = x_A \, \mathbf{1} + y_A \, \mathbf{J},$$

where x_A and y_A are the coordinates of A

$$x_A = L \cos \theta, \quad y_A = L \sin \theta.$$

Hence

$$\mathbf{r}_A = L \cos \theta \, \mathbf{1} + L \sin \theta \, \mathbf{J}. \tag{9.3}$$

The motion of the pendulum is planar, consisting of pure rotation about the pivot
point. The directions of the angular velocity and angular acceleration vectors will be
perpendicular to this plane, in the z direction. The angular velocity of the rod can be
expressed as

$$\boldsymbol{\omega} = \omega \, \mathbf{k} = \frac{d\theta}{dt} \, \mathbf{k} = \dot{\theta} \, \mathbf{k},$$

where ω is the rate of rotation of the rod. This problem involves only a single moving
rigid body and the angular velocity vector refers to that body.

The angular acceleration of the rod can be expressed as

$$\alpha = \alpha \mathbf{k} = \frac{d^2\theta}{dt^2}\mathbf{k} = \ddot{\theta}\mathbf{k}, \tag{9.4}$$

where α is the angular acceleration of the rod. The positive sense is counter-clockwise. The MATLAB program for the kinematics of the rotating link is:

```
% dimension and basic data
L = 0.50;   % length of link (m)
d = 0.01;   % diameter (m)
R = d/2;    % radius (m)
rho = 7800;% density (kg/m^3)
m = pi*R^2*L*rho; % mass (kg)
g = 9.81;   % gravitational acceleration (m/s^2)
alpha = 1.5;
k = 2000; % spring constant
% generalized coordinate theta(t)
syms theta(t)
% angular velocity
omega_ = [0 0 diff(theta(t),t)];
% angular acceleration
alpha_ = diff(omega_,t);
c = cos(theta);
s = sin(theta);
% C center of mass
xC = (L/2)*c;
yC = (L/2)*s;
rC_ = [xC yC 0];
% A end of the link
xA = L*c;
yA = L*s;
rA_ = [xA yA 0];
vA_ = diff(rA_,t);
```

The mass moment of inertia of the rod about the fixed pivot point O can be evaluated from the mass moment of inertia about the mass center C using the transfer theorem. Thus

$$I_O = I_C + m\left(\frac{L}{2}\right)^2 = \frac{m(L^2 + 3\,R^2)}{12} + \frac{mL^2}{4} = \frac{m(4\,L^2 + 3\,R^2)}{12}. \tag{9.5}$$

The MATLAB commands for the mass moment of inertia are:

```
IC = m*(L^2+3*R^2)/12; % mass moment of inertia about C
IO = IC + m*(L/2)^2; % mass moment of inertia about O
```

Two cases will be considered:

(a) the motion of the pendulum before the elastic force is applied and
(b) the motion of the pendulum when the external elastic force is applied.

(a) The force driving the motion of the link is gravity. The weight of the link is acting through its mass center and will cause a moment about the pivot point. This moment will give the rod a tendency to rotate about the pivot point. This moment will be given by the cross product of the vector from the pivot point, O, to the mass center, C, crossed into the weight force $\mathbf{G} = -m\,g\,\mathbf{j}$. Since the rigid body has a fixed point at O the equations of motion state that sum of the moment about the fixed point must be equal to the product of the rod mass moment of inertia about that point and the rod angular acceleration. Thus

$$I_O\,\boldsymbol{\alpha} = \Sigma\,\mathbf{M}_O = \mathbf{r}_C \times \mathbf{G} \qquad (9.6)$$

Substituting Eqs. (9.2), (9.4) and (9.5), the equation of motion for this case is

$$\frac{m(4\,L^2 + 3\,R^2)}{12}\,\ddot{\theta}\,\mathbf{k} = \begin{vmatrix} \mathbf{I} & \mathbf{J} & \mathbf{k} \\ \dfrac{L}{2}\cos\theta & \dfrac{L}{2}\sin\theta & 0 \\ 0 & -m\,g & 0 \end{vmatrix},$$

or

$$\ddot{\theta} = -\frac{6\,L\,g}{4\,L^2 + 3\,R^2}\cos\theta. \qquad (9.7)$$

The initial conditions for this case are

$$\theta(t = 0) = \theta(0) = 0,$$
$$\dot{\theta}(t = 0) = \dot{\theta}(0) = 0.$$

This case ends at the moment $t = t_0$ when $\theta(t_0) = \theta_0$ and $\omega(t_0) = \omega_0$. The equation of motion for the free link is obtained with MATLAB using:

```
G_ = [0 -m*g 0]; % force of gravity at C
MO_ = cross(rC_,G_); % MO = rC_ x G_
% eom for free fall
eq_ = -IO*alpha_ + MO_;
eqz = eq_(3);
```

The second order differential equation is transformed into a system of two first order differential equations with:

```
[V,S] = odeToVectorField(eqz);
syms Y
eom0_ = matlabFunction(V,'vars', {t,Y});
```

The equations of motion are solved using the `event1.m` file:

```
function[value,ist,dir]=event1(~,x)
value = x(1)+pi/6;
ist = 1;
dir = 0;
end
```

The `ode45` function is employed to solve the motion equations:

```
time0_ = [0 1]; % time domain
x0_ = [0 0]; % initial condition
option0 = ...
    odeset('RelTol',1e-3,'MaxStep',1e-3,'Events',@event1);
[tt0, xs0, t0, ye0] = ...
    ode45(eom0_,time0_,x0_,option0);
```

The initial conditions for the contact with the elastic force are calculated with:

```
theta0 = ye0(1);
omega0 = ye0(2);
fprintf('initial conditions for contact\n')
fprintf('theta0 = %6.3g (rad) = %g (degrees)\n',...
    theta0, theta0*180/pi)
fprintf('omega0 = %6.3g (rad/s)\n', omega0)

qt = {diff(theta,t),theta};
list0 = {omega0, theta0};
vAn_ = subs(vA_, qt, list0);
fprintf('vA_ = [%6.3g, %6.3g, %g] (m/s)\n', vAn_)
```

The results are:

```
initial conditions for contact
theta0 = -0.524 (rad) = -30 (degrees)
omega0 =  -5.42 (rad/s)
vA_ = [ -1.36,  -2.35, 0] (m/s)
```

(b) An elastic force with the point of application at A is acting upward on y−axis at t_0,

$$\mathbf{F}_e = k\,|\delta|^{\alpha}\,\mathbf{J},\qquad(9.8)$$

where α is a constant and δ is the deformation given by

$$\delta = L \sin \theta - L \sin \theta_0. \tag{9.9}$$

For this case, the moment with respect to O will be given by the sum of the cross product of the vector from the pivot point, O, to the mass center, C, crossed into the weight force $\mathbf{G} = -m\, g\mathbf{J}$, and the cross product of the vector from the pivot point, O, to the end point of the rod, A, crossed into the elastic force given by the Eq. (9.8):

$$I_O\,\boldsymbol{\alpha} = \Sigma\,\mathbf{M}_O = \mathbf{r}_C \times \mathbf{G} + \mathbf{r}_A \times \mathbf{F}_e. \tag{9.10}$$

Substituting Eqs. (9.3), (9.4), (9.5), and (9.8), the equation of motion for this case is:

$$\frac{m(4\,L^2 + 3\,R^2)}{12}\,\ddot{\theta}\,\mathbf{k} = \begin{vmatrix} \mathbf{I} & \mathbf{J} & \mathbf{k} \\ \dfrac{L}{2}\cos\theta & \dfrac{L}{2}\sin\theta & 0 \\ 0 & -mg & 0 \end{vmatrix} + \begin{vmatrix} \mathbf{I} & \mathbf{J} & \mathbf{k} \\ L\cos\theta & L\sin\theta & 0 \\ 0 & k|L\sin\theta - L\sin\theta_0|^\alpha & 0 \end{vmatrix}$$

or

$$\frac{m(4\,L^2 + 3\,R^2)}{12}\,\ddot{\theta} = -\frac{m\,g\,L}{2}\cos\theta + k\,L\,\cos\theta\,|L\sin\theta - L\sin\theta_0|^\alpha. \tag{9.11}$$

The initial conditions for this case are

$$\theta(t = t_0) = \theta(t_0) = \theta_0,$$
$$\dot{\theta}(t = t_0) = \omega_0.$$

This case will end when θ will reach again the value $\theta(t_f) = \theta_f$ at the moment t_f.
The MATLAB program for the equations of motion is

```
delta = L*sin(theta(t))-L*sin(theta0);
alpha = 1.5;
Fe = k*abs(delta)^alpha; % elastic force
Fe_ = [0, Fe, 0];
% MOe_ = rC_ x G_ + rA_ x Fe_
MOe_ = cross(rC_,G_) + cross(rA_,Fe_);
% eom for elastic contact
eqee_ = -IO*alpha_ + MOe_;
eqze = eqee_(3);
[V,S] = odeToVectorField(eqze);
eome_ = matlabFunction(V,'vars', {t,Y});
```

We start the period of elastic constant at $t = 0$ s. The equations of motion are solved with ode45:

```
timee_ = [0 1];
[tte, xse, te, yee] = ...
    ode45(eome,timee_, [theta0 omega0], option0);
thetaf = yee(1);
omegaf = yee(2);
fprintf('final conditions for contact\n')
fprintf('tf = %6.3g (s) \n',te);
fprintf('thetaf = %6.3g (rad) = %g (degrees)\n',...
    thetaf, thetaf*180/pi)
fprintf('omegaf = %6.3g (rad/s)\n', omegaf)
fprintf('\n')
thetan = xse(:,1);
omegan = xse(:,2);
deltan = L*(sin(thetan) ...
    -sin(theta0)*ones(length(thetan),1));
Fen = k*abs(deltan).^alpha;
thetam = min(thetan);
Fm = max(Fen);
fprintf('thetam = %6.3g (rad) = %6.3g (degrees)\n',...
    thetam, thetam*180/pi)
fprintf('Fm = %6.3g (N)\n', Fm)
```

and the results are:

```
final conditions for contact
tf = 0.0633 (s)
thetaf = -0.524 (rad) = -30 (degrees)
omegaf =   5.42 (rad/s)

thetam = -0.644 (rad) =  -36.9 (degrees)
Fm =   22.5 (N)
```

The link is dropped at a horizontal position $\theta(0) = 0$ and the elastic force $\mathbf{F}_e$ is applied when $\theta(t_0) = \theta_0 = -\pi/6$. At this moment $w(t_0) = w_0 = -5.42$ rad/s. The maximum position of the end A is $\theta_m = -36.9°$. This angle corresponds to the maximum elastic force $F_m = 22.5$ N. The total time of the motion, when the elastic force is acting, has the value $t_f = 0.0633$ s. Figure 9.4 represents the dependence of the angle in time, the variation of the angular velocity, and the elastic force. The animation of the link during the contact with the elastic force is given by:

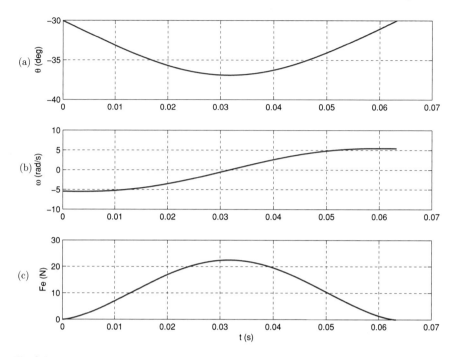

Fig. 9.4 MATLAB simulations: **a** angular displacement, **b** angular velocity, and **c** elastic force

```
for i = 1:10:length(thetan)
clf

xAn = L*cos(thetan(i));
yAn = L*sin(thetan(i));

Ov = zeros(length(thetan(i)),1);
xO = Ov; yO = Ov;

xlabel('x(m)'), ylabel('y(m)')

sf=0.5;
axis([-sf sf -sf sf])
grid on
hold on

% plotting joint O
plot_pin0(0,0,pin_scale_factor,0)

% axis auto
plot([xO xAn],[yO yAn],'k','LineWidth',1.5)
```

```
% label A
textAx = xAn+0.01;
textAy = yAn;
text(textAx, textAy,'A','FontSize',text_size)

% vector Fe
quiver(xAn,yAn,0,.15,0,'Color','r','LineWidth',2.5)
text(xAn+0.01,yAn+.07,'F_e',...
    'fontsize',14,'fontweight','b')

pause(0.2);
end
```

The function `plot_pin0.m` [81] draws a revolute joint linked to the ground as shown in Fig. 9.5.

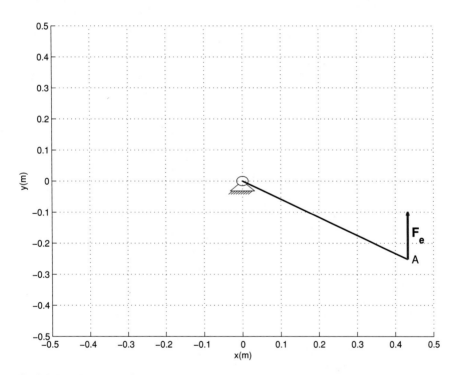

Fig. 9.5 MATLAB plot of the rotating link

The same results are obtained using Lagrange's equations:

$$\frac{d}{dt}\left(\frac{\partial T}{\partial \dot{\theta}}\right) - \frac{\partial T}{\partial \theta} = Q. \tag{9.12}$$

The kinetic energy of the system is:

```
% kinetic energy of the rotating link
% T = 0.5*( m*(vC_*vC_.')+IC*(omega_*omega_.')); or
T = 0.5*IO*(omega_*omega_.');
```

The partial derivative of the kinetic energy, T, with respect to $\dot{\theta}$ is:

```
% dT/d(q')
Tdq = diff(T, diff(theta(t),t));
```

The left hand side of Lagrange's equation of motion is obtained with:

```
% d(dT/d(q'))/dt
Tt = diff(Tdq, t);
% dT/dq
Tq = diff(T, theta(t));
% left hand side of Lagrange's eom
LHS = Tt - Tq;
```

The generalized active force for free fall is

$$Q = \frac{\partial \mathbf{r}_C}{\partial \theta} \cdot \mathbf{G},$$

or with MATLAB:

```
% partial derivative
drC_ = diff(rC_,theta(t));
% force of gravity at C
G_ = [0 -m*g 0];
% generalized active forces Q
Q = drC_*G_.';
```

The Lagrange's equation of motion is solved with:

```
% Lagrange's equation of motion
Lagrange = LHS - Q;
[V,S] = odeToVectorField(Lagrange);
syms Y
eom0_ = matlabFunction(V,'vars', {t,Y});
```

```
time0 = [0 1];
x0 = [0 0]; % initial condition
option0 = ...
    odeset('RelTol',1e-3,'MaxStep',1e-3,'Events',@event1);
[tt0, xs0, t0, ye0] = ...
    ode45(eom0_,time0,x0,option0);
```

For the elastic contact phase the generalized active force is:

$$Q = \frac{\partial \mathbf{r}_C}{\partial \theta} \cdot \mathbf{G} + \frac{\partial \mathbf{r}_A}{\partial \theta} \cdot \mathbf{F}_e,$$

or with MATLAB:

```
delta = L*sin(theta(t))-L*sin(theta0);
alpha = 1.5;
Fe = k*abs(delta)^alpha;
Fe_ = [0, Fe, 0];
% generalized active forces Q
% partial derivative
drA_ = diff(rA_,theta(t));
Q = drC_*G_.'+drA_*Fe_.';
```

The Lagrange's equation of motion is:

```
% Lagrange's equation of motion
Lagrange = LHS - Q;
```

The results are identical with the results obtained with the previous classical method. The complete code for Lagrange method is:

```
clear; clc; close all
% dimension and basic data
L = 0.50;   % length of link (m)
d = 0.01;   % diameter (m)
R = d/2;    % radius (m)
rho = 7800;% density (kg/m^3)
m = pi*R^2*L*rho; % mass (kg)
g = 9.81;   % gravitational acceleration (m/s^2)
alpha = 1.5;
k = 2000; % spring constant
% theta0 = -pi/6;
% generalized coordinate theta(t)
syms theta(t)
```

```
% angular velocity
omega_ = [0 0 diff(theta(t),t)];
% angular acceleration
alpha_ = diff(omega_,t);
c = cos(theta(t));
s = sin(theta(t));
% C center of mass
xC = (L/2)*c;
yC = (L/2)*s;
rC_ = [xC yC 0];
vC_ = diff(rC_,t);
% A end of the link
xA = L*c;
yA = L*s;
rA_ = [xA yA 0];
vA_ = diff(rA_,t);
IC = m*(L^2+3*R^2)/12; % mass moment of inertia about C
IO = IC + m*(L/2)^2;   % mass moment of inertia about O
% kinetic energy of the rotating link
% T = 0.5*( m*(vC_*vC_.')+IC*(omega_*omega_.')); or
T = 0.5*IO*(omega_*omega_.');
% dT/d(q')
Tdq = diff(T, diff(theta(t),t));
% d(dT/d(q'))/dt
Tt = diff(Tdq, t);
% dT/dq
Tq = diff(T, theta(t));
% left hand side of Lagrange's eom
LHS = Tt - Tq;
% partial derivative
drC_ = diff(rC_,theta(t));
% force of gravity at C
G_ = [0 -m*g 0];
% generalized active forces Q
Q = drC_*G_.';
% Lagrange's equation of motion
Lagrange = LHS - Q;
[V,S] = odeToVectorField(Lagrange);
syms Y
eom0_ = matlabFunction(V,'vars', {t,Y});
time0 = [0 1];
x0 = [0 0]; % initial condition
option0 = ...
    odeset('RelTol',1e-3,'MaxStep',1e-3,'Events',@event1);
[tt0, xs0, t0, ye0] = ...
```

```
     ode45(eom0_,time0,x0,option0);
theta0 = ye0(1);
omega0 = ye0(2);
fprintf('initial conditions for contact\n')
fprintf('theta0 = %6.3g (rad) = %g (degrees)\n',...
     theta0, theta0*180/pi)
fprintf('omega0 = %6.3g (rad/s)\n', omega0)
qt = {diff(theta,t),theta};
list0 = {omega0, theta0} ;
vAn_ = subs(vA_, qt, list0);
% double(S) converts the symbolic object S to a numeric object
% converts the symbolic object vBn  to a numeric object
vA_  = double(vAn_);
fprintf('vA_ = [%6.3g, %6.3g, %g] (m/s)\n', vA_)
fprintf('\n');
% elastic contact phase
% elastic displacement
delta = L*sin(theta(t))-L*sin(theta0);
% elastic force
alpha = 1.5;
Fe = k*abs(delta)^alpha;
Fe_ = [0, Fe, 0];
% generalized active forces Q
% partial derivative
drA_ = diff(rA_,theta(t));
Q = drC_*G_.'+drA_*Fe_.';
% Lagrange's equation of motion
Lagrange = LHS - Q;
[V,S] = odeToVectorField(Lagrange);
syms Y
eome_ = matlabFunction(V,'vars', {t,Y});
[tte, xse, te, yee] = ...
     ode45(eome_,time0, [theta0 omega0],option0);
tf = te(2);
thetaf = yee(2);
omegaf = yee(4);
fprintf('final conditions for contact\n')
fprintf('tf = %6.3g (s) \n',tf);
fprintf('thetaf = %6.3g (rad) = %g (degrees)\n',...
     thetaf, thetaf*180/pi)
fprintf('omegaf = %6.3g (rad/s)\n', omegaf)
fprintf('\n')
thetan = real(xse(:,1));
omegan = real(xse(:,2));
deltan = L*(sin(thetan) ...
```

```
      -sin(theta0)*ones(length(thetan),1));
Fen = k*abs(deltan).^alpha;
thetam = min(thetan);
Fm = max(Fen);
fprintf('thetam = %6.3g (rad) = %6.3g (degrees)\n',...
    thetam, thetam*180/pi)
fprintf('Fm = %6.3g (N)\n', Fm)
%plots
figure(1)
subplot(3,1,1)
plot(tte,thetan*180/pi,'LineWidth',1.5)
ylabel('\theta (deg)'), grid
subplot(3,1,2)
plot(tte,omegan,'LineWidth',1.5)
ylabel('\omega (rad/s)'),grid
subplot(3,1,3)
plot(tte,Fen,'LineWidth',1.5)
ylabel('Fe (N)'), xlabel('t (s)')
grid;

function[value,ist,dir]=event1(~,x)
value = x(1)+pi/6;
ist = 1;
dir = 0;
end
```

9.3   Impact of a Free Link with MATLAB

A free bar with the length L, the mass m, and the diameter d is shown in Fig. 9.6. The bar is dropped vertically from rest from a height H and is impacting a solid surface. The MATLAB input data are:

```
L = 0.250;  % length of the bar (m)
d = 0.012;  % diameter of the bar (m)
R = d/2; % radius of the bar (m)
rho = 7800; % density of the material (kg/m^3)
m = pi*R^2*L*rho; % mass of the bar (kg)
g = 9.81; % gravitational acceleration (m/s^2)
% mass moment of inertia about mass center C of the bar
IC = m*(L^2+3*R^2)/12; % (kg m^2)
H = 0.18; % drop height (m)
```

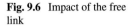

Fig. 9.6 Impact of the free link

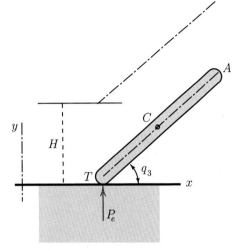

The position of the bar is defined by three generalized coordinates $q_1(t)$, $q_2(t)$, and $q_3(t)$. The impact is initiated when the end T strikes the horizontal surface at $q_2 = 0$. The normal contact force during impact is P_e and the friction force is F_f

$$P_e = k_1 |q_2|^{1.5}, \quad F_f = \mu P_e. \tag{9.13}$$

The linear displacements of the tip, T, of the bar are q_1 and q_2, and the angular position is described by q_3:

```
syms q1(t) q2(t) q3(t)
% q1 x-coordinate of the end T
% q2 y-coordinate of the end T
% q3 angle of the bar with x-axis
```

The angular velocity of the bar is $\omega = \omega \, \mathbf{k} = \dot{q}_3 \, \mathbf{k}$, the angular acceleration is $\alpha = \alpha \mathbf{k} = \ddot{q}_3 \, \mathbf{k}$, and in MATLAB:

```
omega_ = diff([0 0 q3(t)],t); % angular velocity of the bar
alpha_ = diff(omega_,t);      % angular acceleration of the bar
```

The position vector $\mathbf{r}_T$ of the contact point T is

$$\mathbf{r}_T = q_1 \mathbf{1} + q_2 \mathbf{J}.$$

The position vector $\mathbf{r}_C$ of the mass center C of the link is

$$\mathbf{r}_C = \mathbf{r}_T + \mathbf{r}_{TC} = \left(q_1 + 0.5 \, L \, \cos q_3\right) \mathbf{1} + \left(q_2 + 0.5 \, L \, \sin q_3\right) \mathbf{J}.$$

The position vector $\mathbf{r}_A$ of the non-impacting end A of the link is

$$\mathbf{r}_A = \mathbf{r}_T + \mathbf{r}_{TA} = \left(q_1 + L \cos q_3 \right) \boldsymbol{\imath} + \left(q_2 + L \sin q_3 \right) \mathbf{J}.$$

The velocity vector and the acceleration vector of the contact point T are

$$\mathbf{v}_T = \mathbf{P}_T = \dot{q}_1 \boldsymbol{\imath} + \dot{q}_2 \mathbf{J} \quad \text{and} \quad \mathbf{a}_T = \mathbf{P}_T = \ddot{q}_1 \boldsymbol{\imath} + \ddot{q}_2 \mathbf{J}.$$

The position, velocity, and acceleration of the impact point T and of the mass center C are calculated in MATLAB with:

```
% T contact point
rT_ = [q1(t) q2(t) 0];
vT_ = diff(rT_,t);
aT_ = diff(vT_,t);
% C mass center
rC_ = rT_+0.5*[L*cos(q3(t)),L*sin(q3(t)),0];
vC_ = simplify(diff(rC_,t));
aC_ = simplify(diff(vC_,t));
```

The general equations of motion for the planar link for the free fall can be written in the following form

$$m\,\ddot{\mathbf{r}}_C \cdot \boldsymbol{\imath} = \mathbf{0} \cdot \boldsymbol{\imath}, \quad m\,\ddot{\mathbf{r}}_C \cdot \mathbf{J} = \mathbf{G} \cdot \mathbf{J}, \quad I_C\,\boldsymbol{\alpha} \cdot \mathbf{k} = \mathbf{0} \cdot \mathbf{k}. \tag{9.14}$$

The Newton-Euler equations of motion for the free link using MATLAB are:

```
G_ = [0 -m*g 0]; % gravity force
% Newton Euler equations of motion
Neom =-m*aC_+G_;
Eeom =-IC*alpha_;

eqNE1 = Neom(1);
eqNE2 = Neom(2);
eqNE3 = Eeom(3);
```

The equations have to be re-written as a first-order system:

```
[V,S] = odeToVectorField(eqNE1, eqNE2, eqNE3);
% S =
%   q3 -> Y[1]
%  Dq3 -> Y[2]
%   q1 -> Y[3]
%  Dq1 -> Y[4]
%   q2 -> Y[5]
%  Dq2 -> Y[6]
syms Y
```

```
eomNE_ = matlabFunction(V,'vars', {t,Y});
```

The initial conditions for position and velocity of the free fall in MATLAB are:

```
time0 = [0 1];
% initial conditions
% starts from rest
q10=0;    % q1(0)
q20=H;    % q2(0)
q30=pi/4; % q3(0) dropping angle
dq10=0;   % dq1(0)/dt
dq20=0;   % dq2(0)/dt
dq30=0;   % dq3(0)/dt
x0 = [q30 dq30 q10 dq10 q20 dq20];
```

The numerical solutions of the nonlinear system with six ordinary differential equations are solved numerically with an ode45 function. The moment of impact is detected using an event MATLAB command:

```
option0 = odeset...
('RelTol',1e-3,'MaxStep',1e-3,'Events',@eventFL0);

[tt0, xs0, t0, y0] = ...
    ode45(@eomNE_,time0,x0,option0);
```

where eventFL0 is:

```
function [value,isterminal,direction] = eventFL0(~,x)
value = x(5);
isterminal = 1;
direction = 0;
end
```

The free fall ends after t0 = 0.2019s. The initial conditions for the elastic impact are:

```
q3i  = y0(1);
dq3i = y0(2);
q1i  = y0(3);
dq1i = y0(4);
q2i  = y0(5);
dq2i = y0(6);

xi = [q3i dq3i q1i dq1i q2i dq2i];
```

or numerically:

```
% q1  = 0.00 (m)
% dq1 = 0.00 (m/s)
% q2  = 0.00 (m)
% dq2 = -1.88 (m/s)
% q3  = 0.79 (rad)
% dq3 = 0.00 (rad/s)
```

The normal contact force during impact is $P_e = k_1 |q_2|^{1.5}$ and the friction force is $F_f = \mu P_e$. The elastic constant, k_1, is calculated with

$$k_1 = \frac{2 E \sqrt{R}}{3 (1 - \nu^2)}, \qquad (9.15)$$

where $E = 200 (10^9)$ N/m^2 is the Young's modulus and $\nu = 0.29$ is the Poisson ratio. The coefficient of friction is $\mu = 0.1$. The gravity force $\mathbf{G} = -m g \mathbf{J}$ acts at the center of mass, C, of the link and the external impact force acts at the point T. The external force at the contact point is

$$\mathbf{P} = F_f \mathbf{1} + P_e \mathbf{J}. \qquad (9.16)$$

The external force, P_, is calculated with MATLAB as:

```
% elastic impact
E1 = 200e9; % Young's modulus
nu = 0.29; % Poisson ratio
k1 = (2/(3*(1-nu^2))*E1*sqrt(R)); % elastic constant
% fprintf('k1 = %6.3e \n\n',k1)
delta = abs(q2(t));
% elastic normal contact force
Pe = k1*delta^1.5;
mu = 0.1; % coefficient of friction
P_ = [mu*Pe, Pe, 0]; % total elastic contact force
```

The general equations of motion for the free link during the impact can be written in the following form

$$m \ddot{\mathbf{r}}_C = \mathbf{G} + \mathbf{P} \quad \text{and} \quad I_C \alpha = \mathbf{r}_{CT} \times \mathbf{P}, \qquad (9.17)$$

or in MATLAB:

```
% Newton-Euler eom
NeomE =-m*aC_  + G_  + P_;
EeomE =-IC*alpha_ + cross(rT_-rC_, P_);

eqNE1E = NeomE(1);
```

```
eqNE2E = NeomE(2);
eqNE3E = EeomE(3);

[V_,S] = odeToVectorField(eqNE1E, eqNE2E, eqNE3E);
% S =
%
%  q3
% Dq3
%  q1
% Dq1
%  q2
% Dq2
eomI_ = matlabFunction(V_,'vars', {t,Y});
time0 = [0 1];
[tte, xsE, tE, yE] = ...
     ode45(eomI_,time0,xi,option0);

q1E  = yE(3);
dq1E = yE(4);
q2E  = yE(5);
dq2E = yE(6);
q3E  = yE(1);
dq3E = yE(2);
```

At the end of the impact the following results are obtained:

```
% end of impact
% q1  = 0.00 (m)
% dq1 = -1.99 (m/s)
% q2  = 0.00 (m)
% dq2 = 1.88 (m/s)
% q3  = 0.78 (rad)
% dq3 = -24.38 (rad/s)
```

Figure 9.7 shows the vertical elastic displacement q2, the vertical velocity dq2/dt, and the elastic force Fe. The maximum deflection is 6.704e-05 (m).

The kinetic energy of the bar during impact is calculated with:

```
T = 0.5*(m*(vC_*vC_.')+IC*(omega_*omega_.'));
ql = {diff(q1(t),t), diff(q2(t),t), diff(q3(t),t),...
        q1(t), q2(t), q3(t)};
qne = {dq1se, dq2se, dq3se,...
        q1se, q2se, q3se};
Te = subs(T,ql,qne);
```

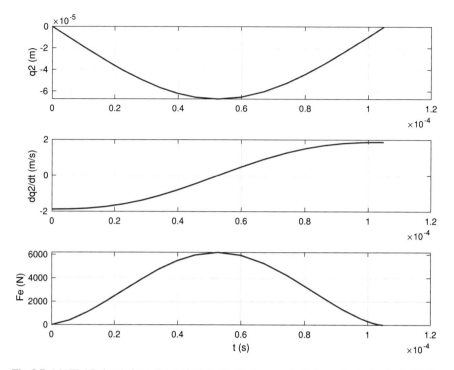

Fig. 9.7 MATLAB simulations: the vertical elastic displacement q2, the vertical velocity dq2/dt, and the elastic force Fe

and the plot of the kinetic energy is shown in Fig. 9.8.

Lagrange's method. The Lagrange's equations of motion for the bar are:

$$\frac{d}{dt}\left(\frac{\partial T}{\partial \dot{q}_i}\right) - \frac{\partial T}{\partial q_i} = Q_i, \quad i = 1, 2, 3. \tag{9.18}$$

The kinetic energy of the system is:

```
% kinetic energy of the bar in general planar motion
T = 0.5*( m*(vC_*vC_.')+IC*(omega_*omega_.'));
```

The partial derivatives of the kinetic energy, T, with respect to $\dot{q}_i$ are:

```
% dT/d(dq/dt)
Tdq1 = diff(T, diff(q1(t),t)); % dT/dq1'
Tdq2 = diff(T, diff(q2(t),t)); % dT/dq2'
Tdq3 = diff(T, diff(q3(t),t)); % dT/dq3'
```

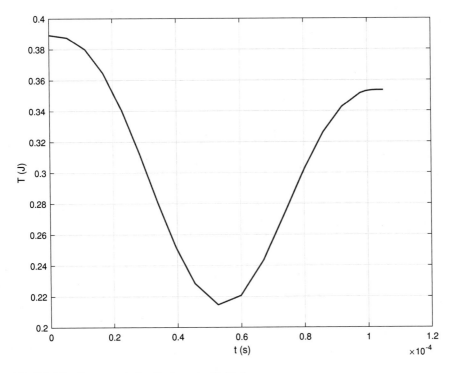

Fig. 9.8 Kinetic energy during the impact with friction

The left hand side of Lagrange's equation of motion for q_1, q_2, q_3 are:

```
% d(dT/d(dq/dt))/dt
Tt1 = diff(Tdq1, t); % d(dT/dq1')/dt
Tt2 = diff(Tdq2, t); % d(dT/dq2')/dt
Tt3 = diff(Tdq3, t); % d(dT/dq3')/dt
% dT/dq
Tq1 = diff(T, q1(t)); % dT/dq1
Tq2 = diff(T, q2(t)); % dT/dq2
Tq3 = diff(T, q3(t)); % dT/dq3
% left hand side of Lagrange's eom
LHS1 = Tt1 - Tq1; % d(dT/dq1')/dt-dT/dq1
LHS2 = Tt2 - Tq2; % d(dT/dq2')/dt-dT/dq2
LHS3 = Tt3 - Tq3; % d(dT/dq3')/dt-dT/dq3
```

The partial derivatives of the position vectors of center of mass, C, and impact point, T, are:

```
% partial diffatives
rC_1_ = diff(rC_, q1(t)); % drC_/dq1
rC_2_ = diff(rC_, q2(t)); % drC_/dq2
rC_3_ = diff(rC_, q3(t)); % drC_/dq3
rT_1_ = diff(rT_, q1(t)); % drT_/dq1
rT_2_ = diff(rT_, q2(t)); % drT_/dq2
rT_3_ = diff(rT_, q3(t)); % drT_/dq3
```

The generalized active forces for the free fall are:

$$Q_i = \frac{\partial \mathbf{r}_C}{\partial q_i} \cdot \mathbf{G}, \quad i = 1, 2, 3,$$

or with MATLAB:

```
% generalized active forces Q for free fall
G_ = [0 -m*g 0];% gravity force
% generalized force corresponding to q1
Q1 = rC_1_*G_.'; % Q1 = G_.drC_/dq1
% generalized force corresponding to q2
Q2 = rC_2_*G_.'; % Q2 = G_.drC_/dq2
% generalized force corresponding to q3
Q3 = rC_3_*G_.'; % Q3 = G_.drC_/dq3
```

The generalized active forces for the impact phase are:

$$Q_i = \frac{\partial \mathbf{r}_C}{\partial q_i} \cdot \mathbf{G} + \frac{\partial \mathbf{r}_T}{\partial q_i} \cdot \mathbf{P}, \quad i = 1, 2, 3,$$

or with MATLAB:

```
% generalized active forces Q for impact phase
% generalized force corresponding to q1
% Q1 = G_.drC_/dq1+P_.drT_/dq1
Q1 = rC_1_*G_.'+rT_1_*P_.';
% generalized force corresponding to q2
% Q2 = G_.drC_/dq2+P_.drT_/dq2
Q2 = rC_2_*G_.'+rT_2_*P_.';
% generalized force corresponding to q3
% Q3 = G_.drC_/dq3+P_.drT_/dq3
Q3 = rC_3_*G_.'+rT_3_*P_.';
```

The Lagrange's equation of motion are obtained with:

```
% first Lagrange's equation of motion
Lagrange1 = LHS1 - Q1;
% second Lagrange's equation of motion
Lagrange2 = LHS2 - Q2;
% third Lagrange's equation of motion
Lagrange3 = LHS3 - Q3;
[V_,S] = odeToVectorField(Lagrange1,Lagrange2,Lagrange3);
S
%   q3
%  Dq3
%   q1
%  Dq1
%   q2
%  Dq2
eomI_ = matlabFunction(V_,'vars', {t,Y});
```

For the free fall the solution is:

```
q10=0;     % q1(0)
q20=H;     % q2(0)
q30=pi/4;  % q3(0) dropping angle
dq10=0;    % dq1(0)/dt
dq20=0;    % dq2(0)/dt
dq30=0;    % dq3(0)/dt
x0 = [q30 dq30 q10 dq10 q20 dq20];
option0 = odeset...
('RelTol',1e-3,'MaxStep',1e-3,'Events',@eventFL0);
% x(5)=q2=0
[tt0, xs0, t0, y0] = ...
     ode45(eomLagra_,time0,x0,option0);
```

For the impact duration the solution is obtained with:

```
q3i  = y0(1);
dq3i = y0(2);
q1i  = y0(3);
dq1i = y0(4);
q2i  = y0(5);
dq2i = y0(6);
xi = [q3i dq3i q1i dq1i q2i dq2i];
time0 = [0 1];
[tte, xsE, tE, yE] = ...
     ode45(eomI_,time0,xi,option0);
```

```
q1E  = yE(3);
dq1E = yE(4);
q2E  = yE(5);
dq2E = yE(6);
q3E  = yE(1);
dq3E = yE(2);
```

The complete code using Lagrange method is:

```
clear; clc; close all
% dimension and basic data
L = 0.250;  % length of the bar (m)
d = 0.012;  % diameter of the bar (m)
R = d/2; % radius of the bar (m)
rho = 7800; % density of the material (kg/m^3)
m = pi*R^2*L*rho; % mass of the bar (kg)
g = 9.81; % gravitational acceleration (m/s^2)
% mass moment of inertia about mass center C of the bar
IC = m*(L^2+3*R^2)/12; % (kg m^2)
H = 0.18; % drop height (m)
syms q1(t) q2(t) q3(t)
% q1 x-coordinate of the end T
% q2 y-coordinate of the end T
% q3 angle of the bar with x-axis
omega_ = diff([0 0 q3(t)],t); % angular velocity of the bar
alpha_ = diff(omega_,t); % angular acceleration of the bar
% T contact point
rT_ = [q1(t) q2(t) 0];
vT_ = diff(rT_,t);
% C mass center
rC_ = rT_+0.5*[L*cos(q3(t)),L*sin(q3(t)),0];
vC_ = simplify(diff(rC_,t));
aC_ = simplify(diff(vC_,t));
% kinetic energy of the bar in general planar motion
T = 0.5*( m*(vC_*vC_.')+IC*(omega_*omega_.'));
% dT/d(dq/dt)
Tdq1 = diff(T, diff(q1(t),t)); % dT/dq1'
Tdq2 = diff(T, diff(q2(t),t)); % dT/dq2'
Tdq3 = diff(T, diff(q3(t),t)); % dT/dq3'
% d(dT/d(dq/dt))/dt
Tt1 = diff(Tdq1, t); % d(dT/dq1')/dt
Tt2 = diff(Tdq2, t); % d(dT/dq2')/dt
Tt3 = diff(Tdq3, t); % d(dT/dq3')/dt
% dT/dq
Tq1 = diff(T, q1(t)); % dT/dq1
```

```
Tq2 = diff(T, q2(t)); % dT/dq2
Tq3 = diff(T, q3(t)); % dT/dq3
% left hand side of Lagrange's eom
LHS1 = Tt1 - Tq1; % d(dT/dq1')/dt-dT/dq1
LHS2 = Tt2 - Tq2; % d(dT/dq2')/dt-dT/dq2
LHS3 = Tt3 - Tq3; % d(dT/dq3')/dt-dT/dq3
% partial diffatives
rC_1_ = diff(rC_, q1(t)); % drC_/dq1
rC_2_ = diff(rC_, q2(t)); % drC_/dq2
rC_3_ = diff(rC_, q3(t)); % drC_/dq3
rT_1_ = diff(rT_, q1(t)); % drT_/dq1
rT_2_ = diff(rT_, q2(t)); % drT_/dq2
rT_3_ = diff(rT_, q3(t)); % drT_/dq3
% generalized active forces Q
G_ = [0 -m*g 0];% gravity force
% generalized force corresponding to q1
Q1 = rC_1_*G_.'; % Q1 = G_.drC_/dq1
% generalized force corresponding to q2
Q2 = rC_2_*G_.'; % Q2 = G_.drC_/dq2
% generalized force corresponding to q3
Q3 = rC_3_*G_.'; % Q3 = G_.drC_/dq3
% Lagrange's eom corresponding to q1
Lagrange1 = LHS1 - Q1;
% Lagrange's eom corresponding to q2
Lagrange2 = LHS2 - Q2;
% Lagrange's eom corresponding to q3
Lagrange3 = LHS3 - Q3;
[V,S] = odeToVectorField(Lagrange1,Lagrange2,Lagrange3);
%   q3 -> Y[1]
%  Dq3 -> Y[2]
%   q1 -> Y[3]
%  Dq1 -> Y[4]
%   q2 -> Y[5]
%  Dq2 -> Y[6]
syms Y
eomLagra_ = matlabFunction(V,'vars', {t,Y});
time0 = [0 1];
% initial conditions
% starts from rest
q10=0;    % q1(0)
q20=H;    % q2(0)
q30=pi/4; % q3(0) dropping angle
dq10=0;   % dq1(0)/dt
dq20=0;   % dq2(0)/dt
dq30=0;   % dq3(0)/dt
```

```
x0 = [q30 dq30 q10 dq10 q20 dq20];
option0 = odeset...
('RelTol',1e-3,'MaxStep',1e-3,'Events',@eventFL0);
% x(5)=q2=0
[tt0, xs0, t0, y0] = ...
    ode45(eomLagra_,time0,x0,option0);
q3i  = y0(1);
dq3i = y0(2);
q1i  = y0(3);
dq1i = y0(4);
q2i  = y0(5);
dq2i = y0(6);
xi = [q3i dq3i q1i dq1i q2i dq2i];
% elastic impact
E1 = 200e9; % Young's modulus
nu = 0.29; % ratio
k1 = (2/(3*(1-nu^2))*E1*sqrt(R)); % elastic constant
% fprintf('k1 = %6.3e \n\n',k1)
delta = abs(q2(t));
% elastic normal contact force
Pe = k1*delta^1.5;
mu = 0.1; % coefficient of friction
P_ = [mu*Pe, Pe, 0]; % total elastic contact force
% generalized active forces Q for impact phase
% generalized force corresponding to q1
% Q1 = G_.drC_/dq1+P_.drT_/dq1
Q1 = rC_1_*G_.'+rT_1_*P_.';
% generalized force corresponding to q2
% Q2 = G_.drC_/dq2+P_.drT_/dq2
Q2 = rC_2_*G_.'+rT_2_*P_.';
% generalized force corresponding to q3
% Q3 = G_.drC_/dq3+P_.drT_/dq3
Q3 = rC_3_*G_.'+rT_3_*P_.';
% Lagrange's eom corresponding to q1
Lagrange1 = LHS1 - Q1;
% Lagrange's eom corresponding to q2
Lagrange2 = LHS2 - Q2;
% Lagrange's eom corresponding to q3
Lagrange3 = LHS3 - Q3;
[V_,S] = odeToVectorField(Lagrange1,Lagrange2,Lagrange3);
eomI_ = matlabFunction(V_,'vars', {t,Y});
time0 = [0 1];
[tte, xsE, tE, yE] = ...
    ode45(eomI_,time0,xi,option0);
q1E  = yE(3);
```

```
dq1E = yE(4);
q2E  = yE(5);
dq2E = yE(6);
q3E  = yE(1);
dq3E = yE(2);
xE = [q1E dq1E q2E dq2E q3E dq3E];
pe = double(subs(Pe, q2(t), q2E));
q1se  = real(xsE(:,3));
dq1se = real(xsE(:,4));
q2se  = real(xsE(:,5));
dq2se = real(xsE(:,6));
q3se  = real(xsE(:,1));
dq3se = real(xsE(:,2));
q2max=max(abs(q2se));
Fen=subs(Pe,q2(t),q2se);
ql = {diff(q1(t),t), diff(q2(t),t), diff(q3(t),t),...
      q1(t), q2(t), q3(t)};
qne = {dq1se, dq2se, dq3se,...
      q1se, q2se, q3se};
Te = subs(T,ql,qne);

figure(1)
subplot(3,1,1)
plot(tte,q2se,'b','LineWidth',1.5),...
ylabel('q2 (m)'),grid,...
subplot(3,1,2)
plot(tte,dq2se,'b','LineWidth',1.5),...
ylabel('dq2/dt (m/s)'),grid;
subplot(3,1,3)
plot(tte,Fen,'b','LineWidth',1.5),
xlabel('t (s)'),ylabel('Fe (N)'),grid;

figure(2)
plot(tte,Te,'b','LineWidth',1.5),...
xlabel('t (s)'),ylabel('T (J)')
grid on

function [value,isterminal,direction] = eventFL0(~,x)
value = x(5);
isterminal = 1;
direction = 0;
end
```

Fig. 9.9 Problem 9.1

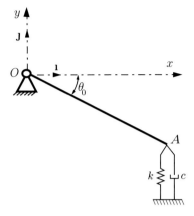

9.4 Problems

9.1 Figure 9.9 shows a slender rotating link of mass m and length L. The link is dropped from rest at a position when $\theta(0) = \pi/4$. The end of the link A is contacting a spring and damper system when the angle, θ, is θ_0. Find the maximum force of the link and the time from the moment when the link is released until the end of the contact with the spring and damper system. Given:

```
L = 0.5;    % length of link (m)
m = 2;      % mass (kg)
g = 9.81;   % gravitational acceleration (m/s^2)
k = 1500;   % spring constant (N/m)
c = 20;     % viscous damping ccoefficient (N s/m)
theta_0 = -pi/4; % angle of contact
```

9.2 The m (kg) slender bar is L (m) in length. A resisting moment about the pin support, O, has the magnitude $b\,d\theta/dt$ (N m) and is opposed to the angular velocity $d\theta/dt$ (rad/s) The resistant moment is caused by the aerodynamic drag on the bar and the friction at the support joint. The pendulum starts from rest at an initial angle θ_0. There are two sets of initial conditions (small and large angles) and two specific instants t_1 and t_2. Given:

```
m = 4; %(kg) mass
L = 2; %(m) length
b = 1.4;
g = 9.81; % gravitational acceleration (m/s^2)
% initial angle theta(0) = theta0
theta01 =  2*pi/180; % (rad) small angle
theta02 = 75*pi/180; % (rad) large angle
% specific instants
```

Fig. 9.10 Problem 9.2

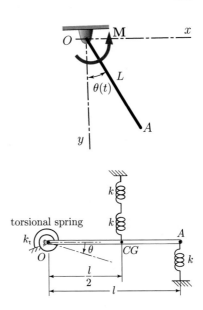

Fig. 9.11 Problem 9.3

```
t1 =  2; %(s)
t2 = 10; %(s)
```

Determine:

(a) response, $\theta(t)$, for small angles and linearized equations of motion at t_1 and t_2.
(b) response, $\theta(t)$, for large angles and nonlinear equations of motion at t_1 and t_2.
(c) plot on the same figure the system responses for linearized and nonlinear equations of motion for for small angles
(d) plot on the same figure the system responses for linearized and nonlinear equations of motion for large angles (Fig. 9.10).

9.3 Find the equation of motion of the rotating uniform rigid bar OA of length l and mass m shown in figure. The gravitational acceleration is g (Fig. 9.11).

References

1. E.A. Avallone, T. Baumeister, A. Sadegh, *Marks' Standard Handbook for Mechanical Engineers*, 11th edn. (McGraw-Hill Education, New York, 2007)
2. I.I. Artobolevski, *Mechanisms in Modern Engineering Design* (MIR, Moscow, 1977)
3. M. Atanasiu, *Mechanics (Mecanica)* (EDP, Bucharest, 1973)
4. H. Baruh, *Analytical Dynamics* (WCB/McGraw-Hill, Boston, 1999)
5. A. Bedford, W. Fowler, *Dynamics* (Addison Wesley, Menlo Park, CA, 1999)

6. A. Bedford, W. Fowler, *Statics* (Addison Wesley, Menlo Park, CA, 1999)
7. F.P. Beer et al., *Vector Mechanics for Engineers: Statics and Dynamics* (McGraw-Hill, New York, NY, 2016)
8. M. Buculei, D. Bagnaru, G. Nanu, D.B. Marghitu, *Computing Methods in the Analysis of the Mechanisms with Bars* (Scrisul Romanesc, Craiova, 1986)
9. M.I. Buculei, *Mechanisms* (University of Craiova Press, Craiova, Romania, 1976)
10. D. Bolcu, S. Rizescu, *Mecanica* (EDP, Bucharest, 2001)
11. J. Billingsley, *Essential of Dynamics and Vibration* (Springer, 2018)
12. R. Budynas, K.J. Nisbett, *Shigley's Mechanical Engineering Design*, 9th edn. (McGraw-Hill, New York, 2013)
13. J.A. Collins, H.R. Busby, G.H. Staab, *Mechanical Design of Machine Elements and Machines*, 2nd Edn. (JWiley, 2009)
14. M. Crespo da Silva, *Intermediate Dynamics for Engineers* (McGraw-Hill, New York, 2004)
15. A.G. Erdman, G.N. Sandor, *Mechanisms Design* (Prentice-Hall, Upper Saddle River, NJ, 1984)
16. M. Dupac, D.B. Marghitu, *Engineering Applications: Analytical and Numerical Calculation with MATLAB* (Wiley, Hoboken, NJ, 2021)
17. A. Ertas, J.C. Jones, *The Engineering Design Process* (Wiley, New York, 1996)
18. F. Freudenstein, An application of Boolean Algebra to the Motion of Epicyclic Drivers. J. Eng. Ind. 176–182 (1971)
19. J.H. Ginsberg, *Advanced Engineering Dynamics* (Cambridge University Press, Cambridge, 1995)
20. H. Goldstein, *Classical Mechanics* (Addison-Wesley, Redwood City, CA, 1989)
21. D.T. Greenwood, *Principles of Dynamics* (Prentice-Hall, Englewood Cliffs, NJ, 1998)
22. A.S. Hall, A.R. Holowenko, H.G. Laughlin, *Schaum's Outline of Machine Design* (McGraw-Hill, New York, 2013)
23. B.G. Hamrock, B. Jacobson, S.R. Schmid, *Fundamentals of Machine Elements* (McGraw-Hill, New York, 1999)
24. R.C. Hibbeler, *Engineering Mechanics: Dynamics* (Prentice Hall, 2010)
25. T.E. Honein, O.M. O'Reilly, On the Gibbs–Appell equations for the dynamics of rigid bodies. J. Appl. Mech. **88**/074501-1 (2021)
26. R.C. Juvinall, K.M. Marshek, *Fundamentals of Machine Component Design*, 5th edn. (Wiley, New York, 2010)
27. T.R. Kane, *Analytical Elements of Mechanics*, vol. 1 (Academic Press, New York, 1959)
28. T.R. Kane, *Analytical Elements of Mechanics*, vol. 2 (Academic Press, New York, 1961)
29. T.R. Kane, D.A. Levinson, The use of Kane's dynamical equations in robotics. MIT Int. J. Robot. Res. **3**, 3–21 (1983)
30. T.R. Kane, P.W. Likins, D.A. Levinson, *Spacecraft Dynamics* (McGraw-Hill, New York, 1983)
31. T.R. Kane, D.A. Levinson, *Dynamics* (McGraw-Hill, New York, 1985)
32. K. Lingaiah, *Machine Design Databook*, 2nd edn. (McGraw-Hill Education, New York, 2003)
33. N.I. Manolescu, F. Kovacs, A. Oranescu, *The Theory of Mechanisms and Machines* (EDP, Bucharest, 1972)
34. D.B. Marghitu, *Mechanical Engineer's Handbook* (Academic Press, San Diego, CA, 2001)
35. D.B. Marghitu, M.J. Crocker, *Analytical Elements of Mechanisms* (Cambridge University Press, Cambridge, 2001)
36. D.B. Marghitu, *Kinematic Chains and Machine Component Design* (Elsevier, Amsterdam, 2005)
37. D.B. Marghitu, *Mechanisms and Robots Analysis with MATLAB* (Springer, New York, NY, 2009)
38. D.B. Marghitu, M. Dupac, *Advanced Dynamics: Analytical and Numerical Calculations with MATLAB* (Springer, New York, NY, 2012)
39. D.B. Marghitu, M. Dupac, H.M. Nels, *Statics with MATLAB* (Springer, New York, NY, 2013)
40. D.B. Marghitu, D. Cojocaru, *Advances in Robot Design and Intelligent Control* (Springer International Publishing, Cham, Switzerland, 2016), pp. 317–325

41. D.J. McGill, W.W. King, *Engineering Mechanics: Statics and an Introduction to Dynamics* (PWS Publishing Company, Boston, 1995)
42. J.L. Meriam, L.G. Kraige, *Engineering Mechanics: Dynamics* (Wiley, New York, 2007)
43. C.R. Mischke, Prediction of Stochastic Endurance strength. Trans. ASME J. Vib. Acoust. Stress Reliab. Des. **109**(1), 113–122 (1987)
44. L. Meirovitch, *Methods of Analytical Dynamics* (Dover, 2003)
45. R.L. Mott, *Machine Elements in Mechanical Design* (Prentice Hall, Upper Saddle River, NJ, 1999)
46. W.A. Nash, *Strength of Materials* (Schaum's Outline Series, McGraw-Hill, New York, 1972)
47. R.L. Norton, *Machine Design* (Prentice-Hall, Upper Saddle River, NJ, 1996)
48. R.L. Norton, *Design of Machinery* (McGraw-Hill, New York, 1999)
49. O.M. O'Reilly, *Intermediate Dynamics for Engineers Newton-Euler and Lagrangian Mechanics* (Cambridge University Press, UK, 2020)
50. O.M. O'Reilly, *Engineering Dynamics: A Primer* (Springer, NY, 2010)
51. W.C. Orthwein, *Machine Component Design* (West Publishing Company, St. Paul, 1990)
52. L.A. Pars, *A Treatise on Analytical Dynamics* (Wiley, New York, 1965)
53. F. Reuleaux, *The Kinematics of Machinery* (Dover, New York, 1963)
54. D. Planchard, M. Planchard, *SolidWorks 2013 Tutorial with Video Instruction* (SDC Publications, 2013)
55. I. Popescu, *Mechanisms* (University of Craiova Press, Craiova, Romania, 1990)
56. I. Popescu, C. Ungureanu, *Structural Synthesis and Kinematics of Mechanisms with Bars* (Universitaria Press, Craiova, Romania, 2000)
57. I. Popescu, L. Luca, M. Cherciu, D.B. Marghitu, *Mechanisms for Generating Mathematical Curves* (Springer Nature, Switzerland, 2020)
58. C.A. Rubin, *The Student Edition of Working Model* (Addison-Wesley Publishing Company, Reading, MA, 1995)
59. J. Ragan, D.B. Marghitu, Impact of a Kinematic link with MATLAB and SolidWorks. Applied Mechanics and Materials **430**, 170–177 (2013)
60. J. Ragan, D.B. Marghitu, MATLAB dynamics of a free link with elastic impact, in *International Conference on Mechanical Engineering, ICOME 2013*, 16–17 May 2013 (Craiova, Romania, 2013)
61. J.C. Samin, P. Fisette, *Symbolic Modeling of Multibody Systems* (Kluwer, 2003)
62. A.A. Shabana, *Computational Dynamics* (Wiley, New York, 2010)
63. I.H. Shames, *Engineering Mechanics—Statics and Dynamics* (Prentice-Hall, Upper Saddle River, NJ, 1997)
64. J.E. Shigley, C.R. Mischke, *Mechanical Engineering Design* (McGraw-Hill, New York, 1989)
65. J.E. Shigley, C.R. Mischke, R.G. Budynas, *Mechanical Engineering Design*, 7th edn. (McGraw-Hill, New York, 2004)
66. J.E. Shigley, J.J. Uicker, *Theory of Machines and Mechanisms* (McGraw-Hill, New York, 1995)
67. D. Smith, *Engineering Computation with MATLAB* (Pearson Education, Upper Saddle River, NJ, 2008)
68. R.W. Soutas-Little, D.J. Inman, *Engineering Mechanics: Statics and Dynamics* (Prentice-Hall, Upper Saddle River, NJ, 1999)
69. J. Sticklen, M.T. Eskil, *An Introduction to Technical Problem Solving with MATLAB* (Great Lakes Press, Wildwood, MO, 2006)
70. A.C. Ugural, *Mechanical Design* (McGraw-Hill, New York, 2004)
71. R. Voinea, D. Voiculescu, V. Ceausu, *Mechanics (Mecanica)* (EDP, Bucharest, 1983)
72. K.J. Waldron, G.L. Kinzel, *Kinematics, Dynamics, and Design of Machinery* (Wiley, New York, 1999)
73. J.H. Williams Jr., *Fundamentals of Applied Dynamics* (Wiley, New York, 1996)
74. C.E. Wilson, J.P. Sadler, *Kinematics and Dynamics of Machinery* (Harper Collins College Publishers, New York, 1991)
75. H.B. Wilson, L.H. Turcotte, D. Halpern, *Advanced Mathematics and Mechanics Applications Using MATLAB* (Chapman & Hall/CRC, 2003)

76. S. Wolfram, *Mathematica* (Wolfram Media/Cambridge University Press, Cambridge, 1999)
77. National Council of Examiners for Engineering and Surveying (NCEES), *Fundamentals of Engineering. Supplied-Reference Handbook* (Clemson, SC, 2001)
78. * * *, *The Theory of Mechanisms and Machines* (Teoria mehanizmov i masin) (Vassaia scola, Minsk, Russia, 1970)
79. eCourses—University of Oklahoma. http://ecourses.ou.edu/home.htm
80. https://www.mathworks.com
81. http://www.eng.auburn.edu/~marghitu/
82. https://www.solidworks.com
83. https://www.wolfram.com

Index

B
Body-fixed reference frame, 16

C
Cam, 189, 190, 192–194
Circular pitch, 165
Contour diagram, 4, 6, 61, 101
Coriolis, 41, 57, 60, 64, 83, 194

D
D'Alembert, 24, 25, 57, 101, 107, 123, 134
Degrees of freedom, 3, 6, 7, 24
Diametral pitch, 165
Direct dynamics, 24, 207
Dyad, 6, 31, 73, 125

E
Elastic force, 207, 208, 211, 215
Epicyclic gear, 165, 167
Equivalent linkages, 198
External moment, 31, 73, 108

F
Follower, 9, 189

G
Generalized active force, 25, 224, 225
Generalized coordinates, 25, 229
Generalized inertia force, 25

I
Independent contours, 1, 5, 170, 181
Inertia force, 24, 87–89
Inertia moment, 24
Initial conditions, 208, 209, 219
Inverse dynamics, 24, 31, 57

J
Joint forces, 46, 101, 150

K
Kane's dynamical equations, 25
Kinematic pair, 1
Kinetic energy, 25, 224, 233, 234

L
Lagrange, 24, 26, 207, 224, 234, 235, 237, 238

M
MatlabFunction, 212
Module, 166, 183
Monoloop, 58
Motor moment, 31, 68, 73

N
Newton-Euler, 24, 46, 47, 73, 91, 93, 207, 230

O
Ode45, 211, 213, 219, 231

© The Editor(s) (if applicable) and The Author(s), under exclusive license to Springer
Nature Switzerland AG 2022
D. B. Marghitu et al., *Mechanical Simulation with MATLAB®*,
Springer Tracts in Mechanical Engineering,
https://doi.org/10.1007/978-3-030-88102-3

Printed in the United States
by Baker & Taylor Publisher Services